The Pursuit of Perfe

The Pursuit of Perfection

Aspects of Biochemical Evolution

Athel Cornish-Bowden

Directeur de Recherche,
CNRS, Marseilles

OXFORD
UNIVERSITY PRESS

OXFORD
UNIVERSITY PRESS

Great Clarendon Street, Oxford OX2 6DP

Oxford University Press is a department of the University of Oxford.
It furthers the University's objective of excellence in research, scholarship,
and education by publishing worldwide in

Oxford New York

Auckland Bangkok Buenos Aires Cape Town Chennai
Dar es Salaam Delhi Hong Kong Istanbul Karachi Kolkata
Kuala Lumpur Madrid Melbourne Mexico City Mumbai Nairobi
São Paulo Shanghai Taipei Tokyo Toronto

Published in the United States
by Oxford University Press Inc., New York

First published 2004

British Library Cataloguing in Publication Data

Data available

Library of Congress Cataloging in Publication Data

Data available

ISBN 0-19-852095-6 (Hbk)

ISBN 0-19-852096-4 (Pbk)

10 9 8 7 6 5 4 3 2 1

Typeset by Newgen Imaging Systems (P) Ltd., Chennai, India
Printed in Great Britain
on acid-free paper by
Biddles Ltd., King's Lynn

To Enrique Meléndez-Hevia, who inspired much of what is written here

Preface

Popular books about evolution are now quite easy to find, and some of them are excellent, but they tend to be books about anatomy or behavior and to ignore biochemistry except insofar as it impinges on these other aspects. Yet in some respects our view of evolution, especially in relation to the questions that interest nonspecialists the most, has been transformed by biochemical information. Thirty years ago all paleontologists agreed that humans had no close relatives in the animal kingdom: they thought that the great apes were far more closely related to one another than any of them was to us, and that they separated from the human line about 30 million years ago, about the same time as both apes and humans diverged from the monkeys. Now nearly all paleontologists agree that the separation between the African apes and humans is far more recent, around 5 million years ago, after the separation of these species from other apes such as orangutans and gibbons, and long after the separation of the apes and monkeys. This is a radical change of opinion. What has caused it? A vast new haul of fossilized bones, perhaps? A series of chimpanzee and human teeth with representatives of every millennium from 30 million years ago until now? No. The supply of hominid and ape fossil bones and teeth is almost as sparse today as it has always been, and it would still be possible to store the entire collection in a small room. No, the change has come from biochemical information, first from protein structures determined in the 1960s, and more recently from gene structures. As new information is accumulating rapidly, and the information available now is a tiny fraction of what will be available in a few years' time, the trend toward basing evolutionary conclusions on biochemical information can only continue.

Deductions based on protein and gene structures are now finding their way into popular books about evolution, but there are other kinds of biochemical information that continue to be largely unknown other than to specialists, and it is with these that I shall be mainly concerned in this book. For example, some aspects of metabolism can be rigorously tested against mathematical criteria to determine whether living systems organize their metabolism in the best way they possibly could, or whether we just see the haphazard result of a series of arbitrary choices of approaches that happened to work but which were not necessarily any better than others. To a considerable degree we find that metabolism is indeed optimized, but biochemists have also been as prone as any other biologists to see adaptation wherever they look, and so we also need to be on our guard against the sort of mistakes that can result from a lack of rigorous logic. We shall see examples of nonsensical adaptation that derive from nonsensical analysis, and of genuinely useful properties of biochemical systems that derive from necessity rather than selection; but we shall also see examples of optimal design in cases where it can only be explained by selection of perfection.

The title of the book is ambiguous: does it refer to the propensity of natural selection to arrive at the best solution to a problem, to the capacity of some biologists to find adaptations wherever they look, or even to the hope of some modern biotechnologists that by tinkering with the genes of microorganisms they can achieve a state of perfection that natural selection has not managed to reach? In truth, it refers to all three, though the first is the most important and was what I had in mind in choosing it.

I owe a great debt to Enrique Meléndez-Hevia, whose name appears several times in the middle part of this book. Until I read his own book *La Evolución del Metabolismo: hacia la Simplicidad* (Evolution of Metabolism: toward Simplicity) I don't think it had ever occurred to me that I would want to write a nontechnical book of my own, and when I started I was mainly interested in trying to bring Enrique's ideas to a broader audience. As the work proceeded it came to encompass not only a broader audience, but also a broader canvas, but his ideas remain very much present in Chapters 4–6, and have been influential in the rest as well.

I also owe a great debt to David Fell, who went through the whole text in detail, first on his own and then with me, and suggested a great many improvements. I am also grateful for other useful comments from the reviewers who read the book on behalf of the publisher, and from Marilú Cárdenas, Peter Duesberg, Jannie Hofmeyr, Enrique Meléndez-Hevia, Chris Pogson, and David Rasnick.

Athel Cornish-Bowden
Marseilles

Contents

1
Some Basic Biochemistry

"Best enzyme man in the world," I said.
I heard Dawlish cough.
"Best what?" he said.
"Enzyme man," I said, "and Hallam would just love *him."*
"Good," said Dawlish.

Len Deighton, *Funeral in Berlin,* 1964

From the elephant to the butyric acid bacterium—it is all the same!

Albert Jan Kluyver, handwritten lecture notes, 1926

Many popular books on evolution are largely books about behavior, and very fascinating they are to read too. Their authors have an easier task than mine, because they can assume that their readers have at least some idea of the basic terminology of their subject. They do not have to spend their earlier chapters explaining what eating is for, what sex involves, how elephants differ from artichokes, and so forth; instead they can march straight into their subject, secure in the knowledge that their readers have at least some idea what they are talking about. Later on, of course, they may want to define some of their terms more precisely, but they can safely leave this task until the reader's attention is well and truly engaged.

Biochemical evolution is more difficult to describe, however, because biochemistry is a less well-known science that does not feature much in popular literature. When you do find a mention of a biochemist in a novel, it is likely to be as vague as the quotation at the head of this chapter, and even reading the whole book that it is taken from will not tell you what an enzyme is, why this may be useful or important to know, or even that the study of enzymes constitutes a part of the study of biochemistry. Likewise, we are introduced quite early in Tom Sharpe's novel *Wilt* to a character described as a biochemist, but a few pages later we learn that he wrote his doctoral thesis on "role play in rats," and nothing in the rest of the book suggests that he ever did anything that a biochemist would recognize as biochemistry. Fortunately, most of the details of biochemical knowledge do not matter much for the purposes of this book, but some of them do, and so I shall begin with a few pages introducing some of the most basic ideas of biochemistry.

Before getting down to any details, there is one point about biochemistry that is so familiar to biochemists that they do not always remember to mention it, though it appears bizarre in the extreme to other kinds of biologists. In any other sort of biology it is essential to be clear about which organism or group of organisms you are talking about, because almost everything in biology varies wildly from one sort of organism to another. Biologists are accordingly rather shocked at the vagueness—to the point of sloppiness—of biochemists about species. To take one example, there is a protein known as hexokinase that is necessary for the first step in obtaining energy from glucose, a major dietary component; it is found in organisms as diverse as yeast, wheat, rats, and bacteria. It happens that the form of hexokinase that has been studied in most detail comes from brewer's yeast, *Saccharomyces cerevisiae*, and when biochemists refer to "hexokinase" without qualifying it, they usually mean yeast hexokinase; even if the discussion is about human metabolism and requires some knowledge of human hexokinase, the biochemist will often fill in gaps in knowledge by assuming that the human is the same as yeast.

This sort of thing can seem quite absurd to a zoologist: how could you guess some unknown fact about human anatomy by looking at the corresponding feature of a fungus? What would be the "corresponding feature" anyway, for creatures so different from one another? Amazingly, however, it works quite well most of the time in biochemistry, because all organisms are made from the same components—proteins, carbohydrates, and fats—use the same sort of chemistry to make these out of the food they consume and convert them into one another, and use the same method for storing the information they need for reproducing themselves. To a first approximation, therefore, biochemistry is indeed the same in all organisms, the point of the remark quoted at the beginning of this chapter from the Dutch microbiologist Albert Kluyver. This is nowadays more familiar in various versions used by Jacques Monod, of which "anything that is true of *Escherichia coli* must be true of elephants, only more so" expresses the idea most forcefully. *Escherichia coli*, a bacterium originally isolated from human feces, is one of the most intensively studied organisms in biochemistry. Much of what we know about metabolism came from studies of it, and biologists have sometimes contemptuously referred to it as "the only animal known to biochemists."

This idea of the unity of biochemistry is not just a lazy shortcut for second-rate scientists unwilling to spend the time filling in all the details: it is taken quite seriously as a way of rationalizing and integrating the subject, and, as noted a moment ago, was popularized by Jacques Monod, the great bacteriologist and biochemist who won the Nobel Prize for Physiology and Medicine in 1965. Along with Francis Crick and James Watson, he is among the few Nobel prizewinners whose names remain familiar some years after the award. Much earlier than him, we can find Geoffroy Saint-Hilaire commenting in 1807 that "there is, philosophically speaking, only a single animal," or Thomas Henry Huxley, in 1868, that "all living forms are fundamentally of one character." Whatever reservations we may have about such statements in biology as a whole, there is no doubt that they have taken biochemistry a long way.

Elephants are not quite the same as *E. coli*, of course, but so far as their biochemistry is concerned the similarities are more striking than the differences: both contain proteins, their proteins are made with the same 20 basic ingredients, the choice between these ingredients is made according to the same genetic code, written in the same genetic language, both base their cellular economy on the extraction of energy from glucose, and so on. The proteins that both contain are, moreover, used for the same sorts of purpose: protein forms the principal building material for both *E. coli* and elephants—hair, skin, and many other structural elements being made of protein. Both *E. coli* and elephants likewise use proteins as their preferred material for making the molecules that allow all the chemical reactions in a living organism to take place. Substances that need to be present for a reaction to be possible—even though they are not themselves consumed or produced by it—are called *catalysts*, and these are used not only in living organisms but also in many industrial processes. Natural catalysts in living organisms are usually called *enzymes*, and it is these that are nearly always made of protein.

Later in this book (in Chapters 4 and 5) we shall be discussing metabolic pathways, the sequences of chemical reactions that interconvert different cellular components. These are not absolutely identical in all organisms, but again, the similarities are very striking. For example, the sequence of reactions that convert sugar into alcohol in yeast fermentation occurs in almost exactly the same form in the human, and it has almost exactly the same function: to extract energy from glucose. Only the end of the process is different, because yeast makes alcohol and we do not. Despite that, the enzyme (alcohol dehydrogenase) that actually makes the alcohol in yeast exists also in ourselves, but the reaction goes in the opposite direction: we use it to remove alcohol.

It is sometimes said in England that even the Archbishop of Canterbury consists of 99% water, and doubtless similar ideas are current in other countries. This isn't strictly true, in fact: unless the Archbishop is extremely far from the average he is only about two-thirds water.[1] Nonetheless, the idea is qualitatively right, as water is by far the most abundant component of the human body, and indeed of almost any other living organism. The chemistry of life, therefore, is largely the chemistry of processes in water. Although we now know that some of the most interesting processes occur in or across the membranes that separate cells or parts of cells from one another, water is usually involved even then.

This may seem obvious, as water is also by far the most familiar fluid in our lives, but it is important to realize that chemistry of reactions in water under mild conditions of acidity and temperature is not what most chemists

[1] The version that was current when I was a child was probably an exaggerated adaptation of a much more accurate remark of the great geneticist J. B. S. Haldane, to the effect that even the Pope was 70% water.

study most of the time, and that what a chemist regards as very mild conditions may be quite extreme from the point of view of a living cell. On the shelves of any (old-fashioned) chemistry laboratory you are likely to find a bottle labeled "dilute HCl," meaning dilute hydrochloric acid. A substance is called an acid if it is a good source of protons, that is, hydrogen atoms that are positively charged because they have lost their negatively charged electrons. A large part of chemistry, especially the chemistry of processes in water, is concerned with moving protons around. Now, what a chemist calls "dilute acid" actually contains a concentration of protons more than 10 million times greater than that in a typical living cell. By these standards, an inhabitant of Greenland would find it quite reasonable to refer to India as sparsely populated.

It follows that the reactions necessary for life must proceed under conditions vastly milder than those considered in chemistry to be mild. As virtually all the reactions we are concerned with are extremely slow under such mild conditions—for practical purposes most of them do not proceed at all—they require catalysts, which in living systems are, as noted a little earlier, enzymes. Until a few exceptions were discovered in recent years they were all believed to be made mainly or entirely of protein. For most biochemical purposes, and specifically for those of this book, we will not go far wrong if we continue to think of enzymes as proteins.

We are usually impressed by the fact that enzymes are very effective catalysts, that is, by the fact that they can make normally very slow reactions proceed very fast. This misses a crucial point, however. Making a reaction go faster is one of the easiest things a chemist can do. Virtually any reaction can be made as fast as you like by confining the reactants in a sealed container and heating them to a very high temperature. There are no important exceptions to this generalization, so why should we be impressed at the capacity of enzymes to do the same thing? The fact that an enzyme can do it at low temperatures is part of the answer, but the really impressive aspect of an enzyme is not that it is a good catalyst for a given reaction but that it is an extremely bad catalyst—no catalyst at all, in fact—for virtually every other reaction. The sealed-tube approach, effective though it is, is just as effective at accelerating large numbers of other reactions that we do not wish to accelerate, in other words it lacks *specificity*. Specificity is what enzymes abundantly provide, and it is their specificity that ought to impress us, because it is specificity that allows all the reactions of life to proceed in an orderly fashion under the mild conditions that exist in a living cell. These mild conditions are not a luxury, nor are they a chance consequence of the fact that we live on a planet with abundant water at a moderate temperature; they are an absolute necessity for life. Without mild conditions it would be impossible to prevent a mass of unwanted reactions from proceeding in parallel with the desirable ones. This raises another question: if mild conditions are essential, how is it that certain organisms, known as "extremophiles," manage to live in conditions far less mild than what we take for granted—at temperatures far above the ordinary boiling point of water near to deep ocean vents, for example? This is an interesting question, but I shall not develop it here.

Specificity comes, however, with a large cost. If the catalysts are to be highly specific there must be many different kinds of catalysts, to a rough estimate one for each reaction that the cell needs to undergo. Even reactions that go quite fast without one may need to be catalyzed in the living organism. For example, removing water from dissolved bicarbonate so that it can be released in the lungs as a gas—carbon dioxide—is a very fast reaction without any catalyst. It requires only a matter of seconds to go almost to completion, but that is not fast enough, and many living organisms use the enzyme carbonic anhydrase to accelerate it by many thousandfold. Some bacteria even use two different carbonic anhydrases on the two sides of a membrane: one to convert the negatively charged bicarbonate ion into neutral carbon dioxide so that it can pass through the membrane as an uncharged molecule, the other to convert it back into bicarbonate on the other side. Specificity also means that enzymes have to have complicated (and hence large) structures: a simple structure such as that of the metal platinum may work well as a general catalyst, but there is no way to make a specific catalyst with anything as simple as platinum.

An enzyme is a large molecule, normally at least 20 000 times greater in mass than a hydrogen atom, and usually much more than that: compare this, for example, with a molecule of the sugar glucose, which is a mere 180 times heavier than a hydrogen atom. Proteins (including enzymes) and other large biological molecules are often called *polymers*, which means that their large structures are achieved by assembling large numbers of much smaller building blocks, these small building blocks being chosen from a fairly small number of possibilities. A protein may contain a small number of identical (or nearly identical) *subunits*, these being very large themselves. In this book I shall largely ignore subunits, and when I talk about the structure of a protein I shall actually in most cases be talking about the structure of one of its subunits.

However, it is hardly satisfying to say that enzymes are large compared with the molecules they act on and leave it at that. Why do they have to be so big? Actually this is partly a misunderstanding, and partly a real question that deserves an answer. Before trying to answer the valid question I shall dispose of the misunderstanding, which comes from the way in which we commonly perceive the sizes of things. When we talk about things we can handle, like beds or bicycles, we normally think in terms of linear dimensions. So, if we say that a bed is a bit bigger than a person we mean, and are understood to mean, that it is a bit longer than the person who sleeps in it. The fact that it is also a bit higher and a lot wider, so that its volume is much larger than that of the person who sleeps in it, goes if not unnoticed then at least unremarked. On the other hand, if we say that a bicycle is about the same size as its rider, we mean, and are understood to mean, that it has about the same length, even though it occupies a much smaller volume.

However, when we talk about the sizes of things that we cannot handle and are far outside the range of everyday objects, like proteins or planets, we are usually referring to mass rather than length. So, noting that an enzyme like liver hexokinase is some seventy times larger (in mass, and also in volume, as

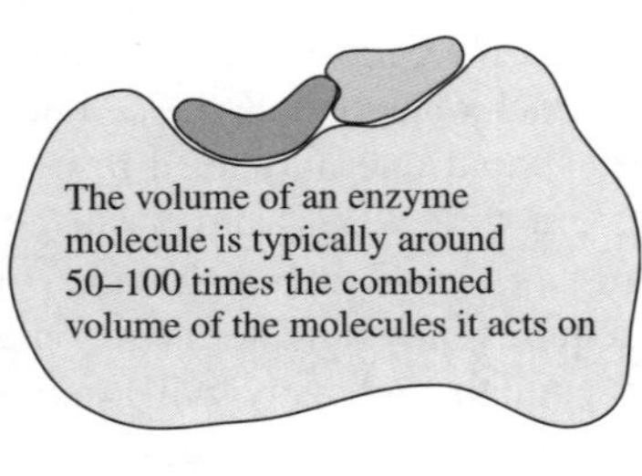

Fig. 1.1 Sizes of familiar and unfamiliar objects. A typical small room used as an office for one person has a volume about 200 times the volume of the person (left). An enzyme is often regarded as a very large molecule even if it occupies no more than 50 times the volume of the molecules it acts on (right)

the densities are not very different) than the combined size of the two molecules that it acts on (commonly called its *substrates*), we tend to be impressed by the large factor, forgetting that its cube root is not much greater than four; the enzyme is actually only a bit more than four times the "size" (as we would use the word for beds or bicycles) of its substrates. In fact a volume ratio of 70 is not far out of line with what we regard as normal in everyday life, such as the size of a car considered suitable for transporting a single commuter to work, and it is far less than many ratios in the macroscopic world: a room only 70 times bigger in volume than the person who worked in it would be regarded as tiny. People who complain of having to work in "boxes" or "cupboards" are usually referring to volume ratios of the order of 100, but they may sometimes be the same people who marvel at how big enzymes are compared with the molecules they act on. This difference between the way we perceive familiar and unfamiliar objects is illustrated in Figure 1.1.

Yet when all is said, a valid question remains. Could not an enzyme like hexokinase have about half the linear size that it has, thus requiring only one-eighth of as much cell resources to synthesize it, while still being large enough to make all of its necessary interactions with its substrates? If it could, then why has evolution failed to produce a smaller and more efficient enzyme? In trying to answer this, we must bear in mind that an enzyme is not just a lump of amorphous glob of roughly the same size as the molecules it works on. If it were, it would be unlikely to do anything much, and certainly nothing precise. An enzyme is a precision instrument, capable of recognizing its substrates, distinguishing them from other molecules of similar size and chemical behavior, and transforming them in precise ways. In general, the more precise any instrument has to be the bigger it must be, and the relative size difference between an enzyme and the substance that it acts on, is not very different from what you find if you compare a precision drill with the object to be drilled, if you include the clamps and other superstructure as part of the drill.

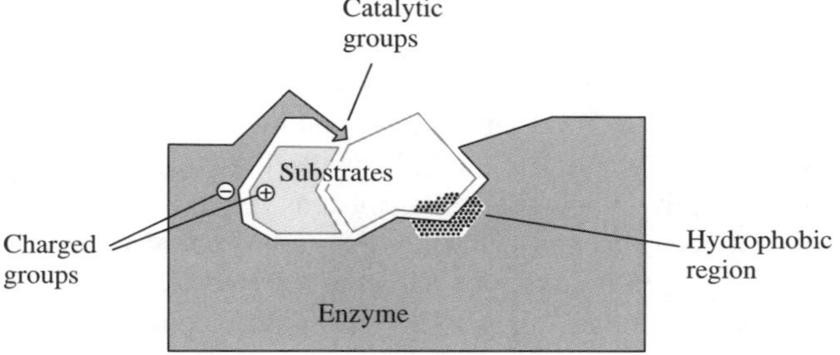

Fig. 1.2 Features necessary for an enzyme to do its job. An enzyme molecule needs a cavity into which its substrates will fit (and into which, as far as possible, unwanted molecules will not fit), and it needs to have regions that attract the substrate molecules, such as charged groups with the opposite charges from groups on the substrates, and hydrophobic (water-repelling) regions to interact with similar regions on the substrates. When all this is achieved it needs to bring its reactive groups into the right positions to bring about the required reaction in the substrates

There is an example in Figure 1.2. It is quite crude, but it does illustrate some of the things that an enzyme must achieve in order to catalyze a reaction. Notice first that not just one substrate is shown, but two, and the enzyme has to promote some reaction between them. This is typical, but not universal: a few enzymes act on just one substrate at a time, but it is much more common to act on two, and some act on three or more. So the enzyme must be able to provide a cavity into which the substrate(s) will fit, and if there are multiple substrates they need to fit side by side with the parts that are to be changed in the reaction close to one another. However, it is not enough just to provide a cavity of the right shape. If a substrate has a charged group (marked + in the example), it will bind much better to a cavity that has an oppositely charge group lined up with it (marked −). Many substrates have *hydrophobic* (water-repelling) regions, and these will bind best to similar water-repelling regions on the enzyme, and so on. However, binding in a cavity is also not enough, as the enzyme needs to do things to the substrates once they are bound, and for this it needs *catalytic groups*, which, again, need to be positioned exactly where they can interact with appropriate parts of the substrates. Moreover, as these catalytic groups may need to move in order to do their jobs, the enzyme will have to have the necessary superstructure to move them in the precise ways needed.

It follows, then, that an enzyme cannot be arbitrarily small, and it must be built according to a structural plan that allows for very precise positioning of its chemically active groups. Moreover, these groups must be extremely unreactive when not in the presence of the molecules they are intended to transform, to make sure that they do not undergo any unwanted reactions, as we saw in discussing specificity. This is much more difficult for working on a very small molecule than it is for a large one, and we sometimes find that the largest enzymes work on the smallest molecules, and vice versa. Enzymes involved in digesting large molecules, such as pepsin in the human stomach, are among the smaller enzymes, whereas catalase, which acts on two molecules of four atoms each, is more than 3000 times their combined weight.

Catalase occurs in blood, and it offers an excellent opportunity to illustrate the tremendous catalytic powers of enzymes, as it acts on a commonly used domestic antiseptic, hydrogen peroxide. Fill a cup about three-quarters full with ordinary hydrogen peroxide of the strength used as an antiseptic (probably labeled as "ten volumes," or as "3%"). Do this in a place where you do not mind if it overflows onto the floor, and then allow a drop of blood to fall into the cup: the result will be instantaneous, as the liquid froths like champagne at the end of a car race, and becomes quite warm to the touch. You may wonder why blood (like other cells from living organisms) needs an enzyme to destroy an ordinary chemical like hydrogen peroxide, and needs it enough to have it available in quite large amounts. The reason is that any living organism needs to be able to destroy harmful chemicals of one kind or another that get into cells. There are a great many potential hazards of this kind, and as the cell cannot prepare a separate solution to every conceivable problem that might arise, it deals with many unwanted chemicals by making them react with ordinary oxygen, in a process known as *oxidation*. This is a useful step, for example, toward making insoluble poisons soluble in water, so that they can be excreted in the urine. The exact nature of the chemical hazard cannot be predicted, however, and so it is not possible to make a separate enzyme to deal with each one. Instead, general-purpose enzymes have evolved, but unfortunately the very fact of being of general purpose makes them unspecific, and they tend to oxidize water instead of the intended substrates, generating hydrogen peroxide. As this is potentially very harmful itself, the cell needs a way of destroying it instantly whenever it appears, and that is what catalase is for.

It turns out that protein molecules fulfill the requirements for a biological catalyst admirably. A protein is composed of a large number of building blocks known as *aminoacids*, arranged head-to-tail in a long line. Each aminoacid has at least one amino group and at least one carboxyl group, and the protein structure involves bonds between the carboxyl group of one and the amino group of the next. Contrary to what is said in most books, the amino group exists normally in a form in which it is very weakly acidic (and is more correctly though in practice almost never called an ammonio group), whereas the carboxyl group normally exists in a form in which it is a very weak base. There are 20 different kinds of aminoacids normally found in proteins (not counting a few that are found in only a few proteins and are inserted into them in special ways), but the order of aminoacids in any protein is neither haphazard nor repetitive. Although a casual glance at the order of aminoacids in any protein might suggest a random or haphazard arrangement, it is not. Some substitutions or rearrangements, especially in the less-important parts of a protein, may be tolerable, but simply rearranging exactly the same building blocks into a random order will no more yield a meaningful protein than rearranging the letters and spaces of a sentence will yield a meaningful sentence: "Novolulogense except iny makeg in bios s tthint of ehe lightion." Even though some of the original structure of the sentence is still there, the modest amount of rearrangement (arbitrary blocks of characters moved from where they were to arbitrary positions, repeated four times) has reduced it to unintelligibility. Incidentally, the sentence appears in its uncorrupted form elsewhere in this book, and as it is a famous sentence in the development of modern biological thought you may be able to recognize it immediately.

Actually this example may be a little misleading, because our eyes and brains are much better at resolving confusion than any machine can be expected to be. Consider, for example, the following sentence, which also appears in an uncorrupted form elsewhere in the book: "As Mark Tawin oerbvsed, while maielnvrlg at our aniamzg aitdtopaan: Our legs are jsut long eoungh to reach the gunord." Here each word has the correct first and last letters, but all the others are in completely random orders. It is striking how much of it can be read almost without effort (though the word "aitdtopaan" may be difficult to identify). However, this is not a good guide to the capabilities of mechanical devices, which lack our capacity to extract some order from disordered information.

The reason why we need as many as 20 different kinds of aminoacids is that a protein needs to have various different kinds of properties, which need to be available at specific locations in space, and the 20 aminoacids differ among one another in the properties they can provide: some are large, some are small; some interact well with water, others repel water; some interact strongly with metals like magnesium, others with metals like zinc; some behave as acids, some as bases; one (histidine) is versatile and can act either as an acid or as a base. We should not be misled by the ending "acid" in "aminoacid," which I am writing as a single word in this book and not as either two words ("amino acid") or a hyphenated word ("amino-acid"), specifically to reduce the risk of this misunderstanding: most aminoacids are no more acidic than they are basic. A free aminoacid is not exactly the same thing as the component derived from it in a protein structure, because it has been chemically changed while making it into a part of a protein, and biochemists often refer to the components of proteins as aminoacid *residues*, that is, as the bits of aminoacids that are left after they have been joined up. However, the distinction is not always made and the word "residue" is sometimes omitted.

It follows that to produce functional proteins a living organism needs a way of specifying the order in which aminoacids are to be arranged, and this order is stored in a special molecule called *deoxyribonucleic acid*, usually shortened to DNA. This is a long sequence of other kinds of building blocks called *bases*, and each protein is specified by a particular stretch of DNA called a *gene*. There are four kinds of bases in DNA, usually known by the first letters of their names, A, C, G, and T, and so a gene can be regarded as a long message written with a four-letter alphabet. Their order is what specifies the order in which aminoacids are to be assembled into a protein. A group of three bases is called a *codon*, and as there are four different kinds of bases there are $4 \times 4 \times 4 = 64$ different kinds of codons. This is more than enough to specify 20 kinds of aminoacids; in fact most aminoacids can be specified in more than one way, but it makes no difference to the resulting protein which particular codon has been used to specify which particular aminoacid. The set of relationships between codons and aminoacids is called the *genetic code*, and it turns out that 61 of the total of 64 codons do in fact code for aminoacids. The other three are "stop" codons, that is, messages telling the protein-synthesizing machinery that it has come to the end of the sequence of a particular protein. The genetic message also needs to include information about where to begin making a particular protein, and how much of it to make, but these are far more complicated aspects that I shall not consider here.

Naively you might expect each gene to be a separate molecule, but it is not so. Instead, genes are strung together one after another in a relatively small number of enormously long DNA molecules. In the human, for example, all of the genes taken together account for a length of a few centimeters, but this is broken up into a mere 46 different molecules. As always when discussing living systems, there is a complication, and in this case it is that the DNA consists not only of genes but also a large amount of "junk"—long stretches of DNA that look at first sight like genes, but which do not in fact code for proteins or anything else. In humans the junk DNA accounts for around 90% of all the DNA, and when it is taken into account the total length of the 46 DNA molecules in a cell is about a meter. Each DNA molecule carries a substantial negative charge, and as negative charges repel one another you might expect the molecule to be extended, and wonder how it is possible to fit a molecule several centimeters in length inside a cell. However, associated with positively charged proteins known as *histones* the DNA makes a compact structure known as a *chromosome*. Packing all this not just into each cell, but into the nucleus of each cell, constitutes a folding problem of tremendous complexity, but it is one that we can largely ignore here: for our purposes it is sufficient to consider a gene as a straight sequence of bases.

It is not quite as sufficient to think of proteins as straight sequences of aminoacids, because understanding their function requires some understanding of how they are folded up into compact globular structures, but it will do for now.

The traditional idea of enzymologists was that life was basically about enzymes and that DNA existed to store the information about how to make them. Much of modern molecular biology has turned this on its head, so that often proteins are just called "gene products," as if what they actually do is of little or no interest. Indeed, much of evolution, at many different levels, can best be understood by accepting the view of Richard Dawkins in his books that a body (let alone a protein) is just the genes' way of perpetuating themselves. However, useful as this may be for understanding why living organisms evolve as they do, it is of little help for understanding the detailed machinery of how living systems work, which is still, indeed, mainly about enzymes.

Enzymes and DNA are not, however, the only important classes of molecules found in living systems. Cells are not just made of proteins and water; and plant cells are not even mainly made of protein and water. There are many kinds of small molecule needed in cells, and in addition to proteins and nucleic acids (the class that contains DNA) together with their building blocks, there are at least two other major classes, the *carbohydrates* and the *lipids*. Carbohydrates include well-known sugars such as sucrose (the everyday "sugar" found on the breakfast table) and glucose, and other sugars that are less well known in everyday life but just as important, such as ribose and deoxyribose, essential components of the genetic material. Glucose itself is a versatile molecule: not only does it provide the main way in which fuel for doing work is available to all the cells of the body, but it is also the major building block for polymers such as cellulose, the primary structural material in plants, and other quite different polymers used for storing energy: starch (in plants) and glycogen (sometimes called "animal starch," in animals). I shall return to glycogen in more detail later in this book (Chapter 6). Lipids include the fats and oils familiar in the diet, as well as other water-repelling substances such as the *phospholipids*, which constitute the major structural component of the membranes that separate cells from one another, and different compartments within cells from one another.

In addition to these we need to mention a small group of metabolites that belong structurally with the building blocks of nucleic acids but which have major metabolic functions that are quite separate from their relationship to nucleic acids. These are the *adenosine phosphates:* two of these, adenosine 5′-triphosphate and adenosine 5′-diphosphate, participate in many metabolic reactions (more, indeed, than any other substance, aside from water); a third, adenosine 5′-monophosphate, participates in relatively few reactions but affects many enzymes as an inhibitor or as an activator. These names are cumbersome for everyday use and biochemists refer to them nearly all of the time as ATP, ADP, and AMP, respectively. In animals, the ATP needed for driving all the functions of the cell is generated in small compartments of cells called *mitochondria*. For the purposes of this book we shall not need to know any details of how mitochondria fulfill their functions, but we do need to know that they exist, because we shall meet them again in a quite different context: it turns out that in most organisms mitochondria contain small amounts of their own DNA, and this allows some special kinds of analyses. Adenosine, the skeleton from which ATP, ADP, and AMP are built, has a separate importance as one of the four bases that define the sequence of DNA.

Before examining them further, we need to pause a moment to consider the branch of science known as *thermodynamics*. The name suggests something to do with heat and power, especially the power to move heavy objects, and this is indeed what thermodynamics is principally about when engineers use it to design power stations or engines. However, thermodynamics has become as important in chemistry, including biochemistry, as it is in engineering, but the reason for its importance there may be less obvious. When engineering was being developed in the nineteenth century, early physicists were trying to understand why some designs of engine worked well whereas others worked badly or not at all. It became more and more clear to them that the principles that decided how energy could be converted into work or different sorts of energy transformed into one another were the same principles regardless of what sorts of energy were under discussion. Thus an electric battery is just as much a machine for converting chemical energy into work as a steam engine is, even though in the case of a battery the conversion is direct: there is no initial conversion of chemical energy into heat followed by conversion of the heat into work. In living organisms muscles behave like motors driven by electric batteries, as they convert chemical energy directly into mechanical work.

In biochemistry, thermodynamics is almost exclusively the study of *equilibria*. When we say that something is thermodynamically possible we mean that the equilibria are such that a process can happen without requiring an external source of energy. This does not, however, mean that it *will* happen, because thermodynamics is not the only factor that decides what will happen. It decides what is allowed and what is not allowed, but no more than that. We see the same thing in the everyday world, in which gravity provides enough energy to bring a large rock from the top of a hill to the bottom, but does not ensure that it will happen: for it to happen, there needs to be a push to get the process started. In biochemistry many reactions are thermodynamically possible but need a catalyst to bring them about.

Returning to ATP and ADP, their adenosine portions are more or less irrelevant from the chemical and thermodynamic point of view, and their job could just as well be done by inorganic triphosphate (three phosphate ions joined together to make a single ion) and diphosphate (two phosphate ions similarly joined together). Indeed, their job is still done by these simpler inorganic molecules in some modern organisms. The important point is that ATP has three phospho groups, one of which can be passed to some other molecule such as glucose, the glucose becoming glucose 6-phosphate and the ATP becoming ADP. Exchanging the diphosphate linkage in ATP for the sugar–phosphate linkage in glucose 6-phosphate is thermodynamically favored, so the process as a whole is thermodynamically favored. The qualification "thermodynamically" may seem cumbersome in this sentence, but it is absolutely necessary: in organic chemistry as a whole and the chemistry of life in particular there is a vast array of chemical reactions that are thermodynamically perfectly possible but which occur extremely slowly in the absence of a suitable catalyst.

ATP is often seen as the universal "cell currency." An energy-harvesting machinery couples the conversion of glucose and other foods to, ultimately, carbon dioxide and water to the resynthesis of ATP from ADP. (Biochemists often refer to this process as *combustion*, in an echo of nineteenth-century ideas of what is involved.) The ATP thus produced is simultaneously reconverted to ADP in a host of metabolic processes that use it to convert metabolites into one another. Although ATP and ADP are themselves metabolites it is often convenient to put them in a separate class, the *coenzymes*, that instead of following a long chain of interconversions are just converted backwards and forwards into one another at many different points in the metabolic network. They are not alone in this category, and several other pairs or small groups of metabolites fulfill similar roles as metabolites: for example, oxidized and reduced NAD (nicotinamide adenosine dinucleotide, but the name does not matter) are similarly converted back and forth in many different metabolic oxidation and reduction reactions. Contrasted with them are the "real" metabolites like glucose, which occur in relatively few distinct steps and form parts of long chains of reactions, or *pathways*.

Some of these metabolites are quite familiar substances, like glucose itself, familiar outside the study of biochemistry because they are mentioned in the lists of ingredients of breakfast cereals and other processed foods, and in leaflets about diet or drugs. Others, like dihydroxyacetone phosphate or 3-phosphoglycerate, may seem rather obscure. As it will often not be necessary to define their structures at all in order to understand their role in the discussion, I shall not do so, or I shall confine the definition to a statement about how many carbon atoms they contain. In general, you may assume that if I refer to a series of three or four reactions that transform the familiar molecule glucose into the unfamiliar one dihydroxyacetone phosphate, then dihydroxyacetone phosphate must retain enough of the structure of glucose to be made from it in a few steps, and this is all you will need to know about its structure. This is because metabolic reactions always proceed in simple steps, so that you can never find a single reaction that rearranges dozens of atoms all at once, any more than evolutionary biologists propose the transformation of an ameba into an elephant in one generation.

In fact, dihydroxyacetone phosphate is a much simpler molecule than glucose, one of the few important molecules in biochemistry that looks exactly the same as its image in a mirror, so do not be put off by the long name. Shorter does not necessarily mean simpler in biochemical names, any more than it does with names of places. Los Angeles is said to have started life as El Pueblo de Nuestra Señora la Reina de los Ángeles de Porciúncula: maybe, but it needed a simpler name when it became an important city. Only places you do not need to mention very often can afford the luxury of names like Ruyton of the Eleven Towns (a small village in Shropshire). So, as dihydroxyacetone phosphate does not feature in lists of cereal ingredients, and is not (yet) considered by dietary advisers to be the secret of a long and healthy life, it continues to be known by its chemical name, whereas with more fashionable molecules like folic acid and riboflavin the simple names conceal much more complicated structures.

All of this multitude of components of even the simplest cell implies a large number of different reactions to connect them all, and a correspondingly large number of enzymes to catalyze them all. The whole network of reactions is called *metabolism*, and the core of this book is about metabolism. It has been a major component of teaching biochemistry to medical and life-science students, and has typically been taught as if the whole complicated organization was arbitrary or haphazard, the result of a whole series of accidents—chance solutions to problems, no better or worse than other solutions that happen not to have been adopted—over the long period of evolution since the origin of life some thousands of millions of years ago. I shall try to convince you that this view is wrong, because although the amount of metabolism that has been analyzed from the point of view advocated here is only a tiny proportion of the whole of metabolism, it leads ineluctably to the conclusion that the solution adopted is never just one of a series of equally good or bad solutions that could have been chosen, but it is always the best solution.

Before trying to convince you of this, however, I need to examine what is involved in protein evolution, and the next chapter is devoted to this. In the one after I shall describe a point of view that is called Panglossian, so that you can approach the rest of the book in a proper spirit of skepticism and you can judge whether I am seeing optimal solutions to metabolic problems because they really are optimal, or because that is what I want to see.

2
The Nuts and Bolts of Evolution

Mechanisms and molecules have been preserved in bacteria, fungi, plants, and animals, essentially intact through billions of years of Darwinian evolution.

Arthur Kornberg, 2000

As we have seen in Chapter 1, a protein is, to the first approximation, a straight string of aminoacids coded by the sequence of bases in a gene. After the essential relationship between DNA sequence and protein sequence was recognized from the famous work of James Watson and Francis Crick in the 1950s, and the code itself was worked out in the 1960s and 1970s, it appeared for a while to be absolutely identical in all organisms and all cells, and it was called the *universal code*. As I shall discuss in the next chapter, we now know that this was an oversimplification, but for the purposes of this chapter I shall ignore the small variations in the genetic code, and treat it as if it really were universal. I shall also ignore other complications in the way the code is read, in particular the noncoding regions, which can in some organisms, including ourselves, represent a very high proportion of the total DNA.

Charles Darwin devoted many pages of *The Origin of Species* to discussing the variations between the individual members of any species. His aim in the book was to convince his readers that evolution was a natural process, which he called *natural selection*, similar to the way in which a horse breeder or a pigeon fancier breeds each new generation by choosing as parents those individuals that come closest to displaying whatever characters are considered desirable. However, before he could convince them of this, he first needed to show that a sufficient amount of variation already exists in a population to allow the necessary selection. Although Darwin knew nothing of genes or their role in coding for proteins, we can now say that the variation he was discussing was due to the existence of multiple variants of genes, with the result that different individuals have different collections of proteins. In biochemical terms, therefore, evolution is the result of changes in genes.

Perhaps the first question we should ask is why any gene changes at all. Would it not be better from the point of view of the gene itself to be copied from one generation to another without any changes whatsoever? Yes, it would, but we are dealing with mechanisms that depend for their efficiency on the laws of chemistry. The genetic code "works" for three principal reasons. First, the double-stranded structure of DNA is more stable if every A is lined up with a T in the other strand, every C is lined up with a G, every G is lined up with a C, and every T is lined up with an A. This allows an exact duplication when the two strands are separated and each strand is then allowed to build a new partner strand by making the correct pairings from a pool of bases. Second, when the gene is translated into protein, each codon is recognized by the enzyme responsible for adding a specific aminoacid to the growing protein. Third, the enzyme that adds any aminoacid can recognize that aminoacid in a mixture with all the other 19.

All of these three recognition steps are inevitably subject to mistakes. Even though it is true that a G lines up better with a C than with a T, and makes stronger interactions with a C, the wrong pairing is by no means impossible (Figure 2.1). Pairing an A with another A or with a G is more difficult, because there is not enough room for the larger bases without distorting the whole structure, but again, it is not utterly impossible. The four bases are not infinitely large and the interactions between them do not involve infinite numbers of chemical groups. Not only are they not infinite, they are not even very large: the chemical energy of a whole pairing can be obtained by adding up the

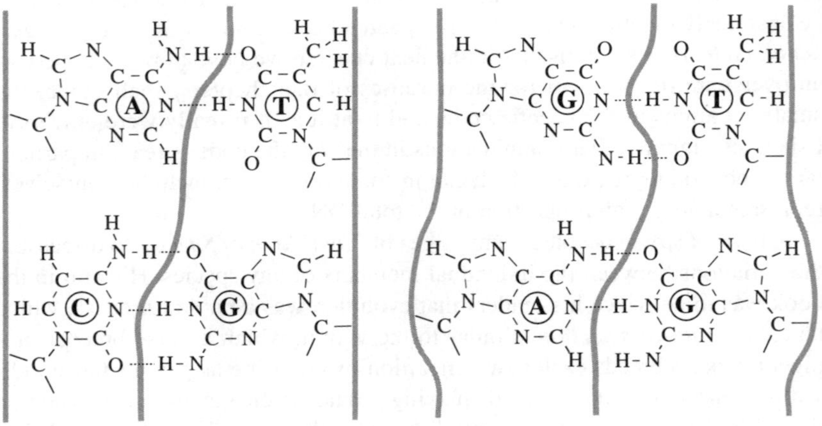

Fig. 2.1 Base pairs in DNA. When the bases are paired correctly they form strong interactions, and the width of the combination is almost exactly the same for each of the four possible combinations (T–A and G–T as well as the two shown), so there is no distortion of the double helix (left). Nonetheless, it is not utterly impossible to form "wrong" base pairs, held together by weaker interactions than the correct pairs (though not infinitely weaker), and when these produce a pair of about the right width, as with the G–T pair illustrated, they can occur fairly easily (right). On the other hand, A–G and several other pairs do not have the right widths and would introduce obvious distortions in the whole structure if they occurred

contributions from the individual groups involved; when this is done the differences between the energy of the correct pairing and the wrong ones are not sufficient to exclude all possibility of a wrong pairing.

It is like putting a left-handed glove on the right hand: not comfortable, but not so difficult as to be impossible to be done accidentally or never to be done deliberately. By calculating the energies involved from chemical considerations and then using different chemical principles to assess how frequent wrong pairings are likely to be in DNA you can calculate a frequency of about one in a million base pairs. This may seem very good—it is certainly much less likely than accidentally putting a left-handed glove on the right hand—but it is not nearly good *enough*, because the huge number of base pairs in all the genes in an organism means that even an error rate as low as this would add up to an intolerable number of errors. For example, in the human, with about 3.5 billion[1] base pairs in the whole genome, an error rate of one in a million implies about 3500 mistakes every time the DNA is duplicated. To prevent this from happening, the machinery for duplicating DNA includes sophisticated mechanisms for detecting and correcting mistakes. In the *germ-line DNA*, which is the DNA that is passed on to the next generation, these mechanisms achieve a final error rate of the order of one in a billion, so the average number of mistakes passed on to the new generation is of the order of three or four. In the cells that are used in the lifetime of the individual but are not passed on to descendants, which are called *somatic cells*, the error rate is higher, but still impressively small.

You may wonder why the somatic cells tolerate a higher error rate than that found in germ-line DNA. The answer is that error detection and correction come at a price. As we have seen, the simple interaction energies in base pairs are not enough to ensure that errors never occur, and detecting them after they have occurred consumes energy: ultimately every base pair checked implies the consumption of a certain quantity of glucose, a small quantity, certainly, but glucose that could otherwise have been used for something else.

The body has to make the same compromises that a factory making a specialized and complicated instrument has to make: the master copy of the design needs to be carefully protected from damage, and if copied for use in another factory it needs to be copied with meticulous accuracy to avoid errors. The more routine copies that are used for the actual manufacturing need to be accurate as well, as it is not good to produce large quantities of defective instruments, but a certain low proportion can be tolerated.

Now that the original designs for most everyday objects are stored in computers, it is no longer easy to find an example from everyday life to illustrate this idea. Before vinyl gramophone records disappeared from the market in the

[1] In this book I follow the usual modern practice of using an American billion, that is, 1 000 000 000, or 10^9. The traditional British billion, a thousand times larger, is a variant that is losing in the struggle for survival that natural selection represents, and in another generation or so may be forgotten by all but historians of language.

face of the onslaught from compact disks they made an excellent example, but I am writing this book 20 years too late. The old principles still probably apply to commercial films for the cinema, however. Somewhere there is a master copy of every film as it was completed in the studio and approved by the producer and director. However, this copy is much too precious to be actually used for showing the film in a cinema. Instead, a number of sub-masters are made from the master copy, as accurately as possible, and sent to various distribution centers, and the master copy is stored away in a vault, protected as far as possible from wear and tear, and if possible never looked at again. The sub-masters are not used in actual cinemas either, but are used to make distribution copies. The distribution copies are then used, and the sub-masters carefully stored. No errors can be tolerated in the master copy, because any error in the master copy will be copied everywhere else. Errors in the sub-masters are marginally less harmful, because they only affect a subset of copies made for use in cinemas, but a high degree of accuracy is still desirable. For the distribution copies, however, faults such as scratches and even missing frames can often be tolerated, because unless they are very severe and frequent they are likely to pass unnoticed.

The same sort of considerations apply to all industrial processes, and, whatever the final product, an economic decision has to be made as to the point at which greater expenditure on error avoidance costs more than the losses incurred by sending a few defective items to market. The place where a manufacturer decides to put this point is a major factor in determining whether a product is considered suitable for the mass market or for users who accept only the best: a top-of-the-range car may cost 10 times as much to buy as a mass-market car of similar dimensions, but it certainly does not use raw materials that cost 10 times as much, and the actual manufacturing costs are unlikely to be 10 times greater either; on the other hand the amount of time, effort, and money devoted to checking for errors and discarding substandard items may be much more than 10 times greater.

As mentioned above, there are other points where accuracy is needed, but these are less important from the evolutionary point of view. Occasional protein molecules that cannot fulfill their functions because they have incorrect aminoacids inserted in some positions represent a waste of resources for the individuals that possess them, but they do not get passed on to the next generation, so they do not accumulate over evolutionary time. Uncorrected errors in DNA replication, however, do get passed on and do accumulate during evolution. Ultimately, they are what evolution consists of. Every single genetic difference that exists between one present-day organism and another—whether we are talking about two different members of the same species, or about two individuals as different as a rhododendron and a tiger—has its origin in a mutation, that is to say an error in the replication of the genetic information.

There are, however, two stages in the process, of which the second is often forgotten in elementary discussions. A mutation is a necessary beginning, but it is not sufficient. If a mutation occurs in an organism that is never born, whether because the mutation itself is lethal or for any other reason, it will never be manifested in any living creature. Even if the individual is born and reaches maturity, the mutation can only affect evolution if there are descendants. In many cases even then, the line containing the mutation may die out in a few generations, so although it may affect a few individuals it never becomes part of what distinguishes one species from another. To do this, the mutation must spread through the entire population—a process known as *fixation*.

It is tempting to think that a favorable mutation will inevitably become fixed, but this is far from being the case, because in the first few generations, when it can only be present in a very small number of individuals, there are many other factors that may cause those individuals to leave no descendants in the long term. Even a highly favorable mutation, say one that increases fitness (loosely the probability of having offspring) by 1%, is almost exactly as likely to be eliminated in the first few generations, and lost forever, as a mutation that has no effect at all on fitness, or a mildly unfavorable one that decreases the fitness by 1%. In these early stages the role of chance is too great for success to be guaranteed. Consider the simplest case, a man with a highly favorable mutation in a gene on his Y chromosome. As we have seen, human genes are organized into 46 chromosomes, but one of these, the *Y chromosome*, occurs only in males (females have two X chromosomes instead), so any mutation in the Y chromosome is inherited only by sons, not by daughters. If the man with such a favorable mutation has no sons then, no matter how many daughters he may have, the mutation will be lost. Even if the favorable mutation occurs on another chromosome, there is a one-half chance that it will not be passed on to any given child. If there are two children, there is a one-quarter chance that neither will receive it, and so on. Even if there are 10 children there is about one chance in a thousand that none of them receive the favorable mutation. This chance may be small, but it is not zero.

The idea of gene fixation is illustrated in the trees of descent in Figure 2.2. To avoid various complications, we consider only people of the same sex who survive to become adults, we treat the generations as completely separate, all members of the same generation reaching maturity at the same time, and we assume a constant population size. With these conditions, a woman has an average of exactly one daughter, though any particular woman may have no daughters, or more than one. The left-hand tree shows that out of 20 women present in generation zero, only 10 had daughters that survived to maturity, and only seven had surviving granddaughters. Thus 13 out of 20 lines have disappeared in just two generations. This does not imply that the 13 potential ancestors were inferior in any way; they were just unlucky. In constructing the tree I did not assume that any women were any more or less likely to have surviving daughters than others. I just allowed each the same probabilities of having zero, one, two, or three daughters. But it is evident that if the population size remains constant some can have three daughters only if others have none.

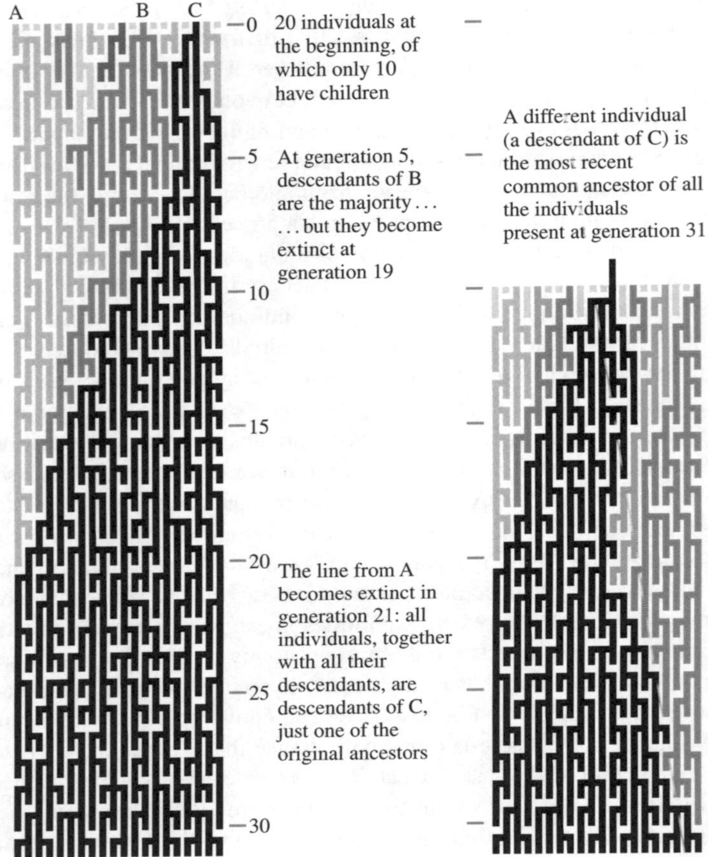

Fig. 2.2 Fixation of genes. The top line of the left-hand tree represents 20 individuals, three of them labeled A, B, and C. These three, and seven others, have children, and 10 do not. Any unique genes possessed by the 10 who leave no descendants must disappear from the population. Although the descendants of B constitute the majority after a few generations they eventually become extinct, and at the bottom of the tree only descendants of C are left. The right-hand tree is identical in structure to the left-hand tree, but only the later generations are shown. Regardless of where one begins, the descendants of all but one of the individuals present in any generation will eventually become extinct

If you follow the black part of the tree down it you can see that in generation 21 only the black line continues; all of the other 19 have disappeared. From this point on, the subsequent generations are all labeled black, and the genetic composition of the individual ancestor under the letter C at the top has become fixed in the population. No further change is possible unless there are mutations. Notice that even with a small population size fixation requires a considerable number of generations. With a larger population size the number of generations required is correspondingly larger, and is in general of the same order of magnitude as the population size. For our own species, as well as for many others in nature, the population size (some billions for humans) is much larger than the number of generations since it separated from other species (hundreds of thousands for humans). This means that there has not been nearly enough time for a significant proportion of genes that appeared after the separation to become fixed. Of course, the human population was very much smaller only a few generations ago, but it has been many generations since it was of the order of hundreds of thousands.

Notice that the illustration can be applied to a single gene, or to the descent from a single individual, or for anything in between. It can thus be applied to the hunt for *mitochondrial Eve*, sometimes regarded as the first woman, though that is an incorrect interpretation, as we shall see. The qualification "mitochondrial" refers to the fact that although both males and females have mitochondria, the ATP-generating compartments in cells that were briefly mentioned in Chapter 1, those in males are not passed on to the next generation. Moreover, although most of the enzymes that a mitochondrion needs are coded in the same DNA as all of the other enzymes in the organism, the mitochondrion contains a small amount of DNA of its own, which codes for a few of its enzymes. All people inherit their mitochondria exclusively from their mothers, and comparing the mitochondrial genes of present-day individuals allows us to deduce something about the mitochondrion that was the most recent common ancestor of all modern ones. Mitochondrial Eve was thus the woman who was the most recent female common ancestor of all living humans.

It is sometimes assumed that mitochondrial Eve was an especially successful individual, probably one with many children. However, it should be clear from the illustration that she was just lucky, and may have had no more than two daughters. (She must have had at least two, because if she had just one then that one would be mitochondrial Eve herself). In the example, the descendants of the ancestor labeled B initially multiplied faster than any other, constituting the majority of all descendants at generation 10, so B seemed a likely candidate to become mitochondrial Eve, but in fact her line disappeared in generation 19, and by generation 21 only descendants of C were left.

As Figure 2.2 shows, mitochondrial Eve should not be confused with the Eve of the Bible, who is normally regarded as the only woman in existence at the time she was created. On the contrary, mitochondrial Eve was just one of many women alive at the same time. The human population was certainly smaller then than it is now, but it was still very large. Because we live in an age of rapid population growth, it is easy to forget that it has not always been so, and that it will not remain so in the future. On the contrary, during most of human evolution, the population size has grown so slowly that a model that treats the population size as constant is not grossly inaccurate.

Moreover, mitochondrial Eve does not remain always the same person. If the left-hand tree represents the descent from a group of 20 women, then the person under the letter C would be mitochondrial Eve so far as the people at generation 21 are concerned, but counting back from the generation at the bottom line shows a much more recent mitochondrial Eve. This is shown in the tree at the right, where the early generations have been omitted, though the late ones are identical to those in the original tree, except that a new shading scheme is used to distinguish the individuals in the new top line at generation 10. There is a new mitochondrial Eve in this tree who is a distant descendant of the first one.

The title of mitochondrial Eve thus advances, and a woman living today or one of her descendants will one day be the mitochondrial Eve for our descendants. She is impossible to identify today, but she must exist. It follows, then, that mitochondrial Eve was not a particularly special person, and did not live at a special moment in human evolution.

Much the same can be said of *Y-chromosome Adam*, the most recent common male ancestor of all men living today. Just as mitochondria descend through the female line, Y chromosomes descend only through the male line. It is sometimes supposed that Y-chromosome Adam and mitochondrial Eve were a couple living at the same time, but not only is there no reason for expecting this to be true, there are very good reasons for thinking it not to be true. Although the average number of children that a man has is the same as the average that a woman has (given that every child has exactly one father and exactly one mother) the variation about the average is not the same for the two sexes. In most mammalian species (including humans) the number of daughters females have is fairly uniform, especially if we just count the daughters who survive to become adults. Except in rapidly growing populations (which can only exist for periods that are very short on an evolutionary scale), a typical woman will have zero, one, two, or three surviving daughters, rarely more. A prolific man, however, may have many more than three surviving sons, and large numbers of men (me, for example) have no sons at all. In some species, like elephant seals, the variations in fecundity between males are greater than they are in humans; in others, such as gibbons, they may be less, but in most mammalian species males are less uniform than females in the numbers of offspring they have.

This may be illustrated by redrawing the tree according to the same principles as Figure 2.2, except that now most individuals have zero, three, or four children, as on the right in Figure 2.3 (the left-hand part being the same as in Figure 2.2 with generations after fixation omitted). Notice that fixation occurs much more quickly, with four distinct fixation sequences in 31 generations. From this sort of argument we may expect that Y-chromosome Adam lived much more recently than mitochondrial Eve, so it is very unlikely that they constituted a couple.

The sort of mutations that I have been mainly discussing, resulting from wrong pairings at specific points in the DNA, and producing individual wrong bases in the daughter DNA, are called *point mutations*. They account for most of the changes that occur when protein sequences change over the course of evolution with little or no change in function. For example, an enzyme called hexokinase D exists in the livers of both humans and rats, where it fulfills a role

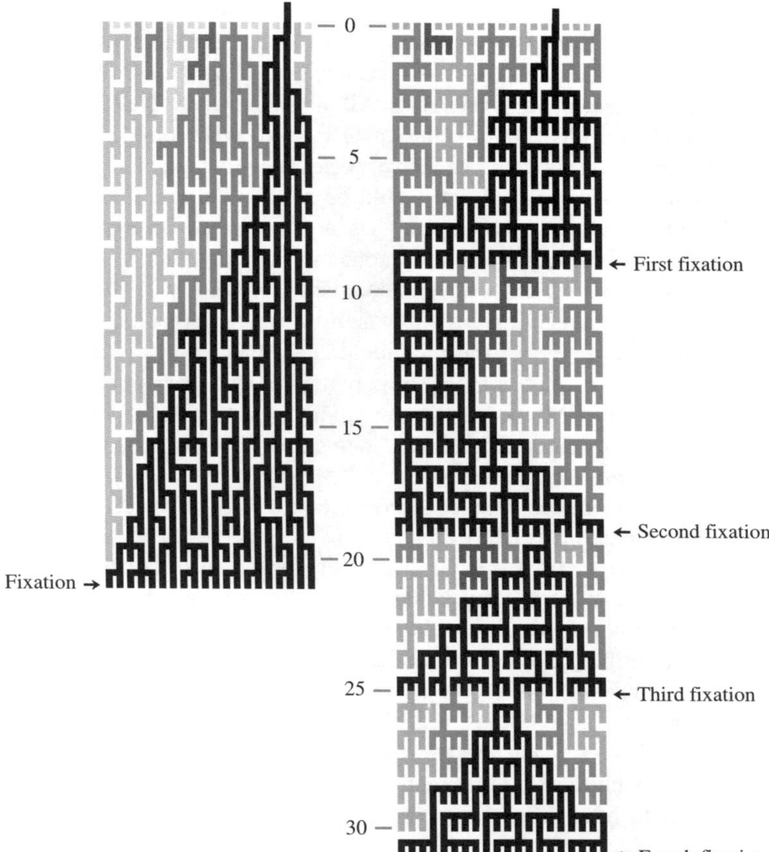

Fig. 2.3 Fixation of genes with large variations in fecundity. The left-hand tree is exactly the same as the left-hand tree of Figure 2.2, except that the generations after fixation are omitted. The right-hand tree shows that when individuals vary greatly in fecundity the lines from individuals of low fecundity become extinct very fast, and fixation requires fewer generations

in controlling the concentration of glucose in the blood, as we shall consider in Chapter 10. The two proteins have more than 460 aminoacids each but are almost exactly the same, with no more than 20 differences between them. Most of these differences are of the simple wrong-substitution kind we have considered, but there are also a couple of *deletions*, as the rat enzyme is a little shorter than the human. However, in properties the two enzymes are more or less indistinguishable, and it is unlikely that a rat would be any the worse off if it had human hexokinase D, or that a human would be any the worse off with rat hexokinase D. We cannot be certain of this, of course, but I do not know of any biochemist who would claim that all variations between protein sequences in different organisms are adaptive.

We find the same sort of results with many proteins. Some, such as hemoglobin, the protein that carries oxygen in our blood, vary much more between species than hexokinase D does; others, such as cytochrome *c*, a protein involved in energy management, vary much less. All of this is very useful for constructing trees of relationship from protein sequences, and cytochrome *c*, for example, was used many years ago to relate many species spanning the plant, animal, and fungus kingdoms—something that would be hard to do with the use of morphology and fossils alone! However, it is not much help in explaining how species become so different from one another during evolution.

Most of the changes that we see in comparing proteins with the same function in different organisms are probably neutral, meaning that they do not have any effect on the fitness of the individual. The realization that most of the mutations detected in such comparisons are neutral was the insight of the great population geneticist Motoo Kimura, and, although there is still some resistance from biologists determined to see adaptation wherever they look, there is increasing acceptance of his view.

This means, therefore, that we need to look elsewhere for an explanation of where new functions come from that allow one species to be different from another. All of the error-correcting machinery works in the direction of forbidding any changes at all, and hence any evolution at all. Such changes as occur are the result of mistakes. Moreover, if Kimura is right then most point mutations will either be lethal, so the individual will not survive, or they will be neutral, so they will have no gross effect. Yet evolution does produce gross effects: a domestic cat is not just a small tiger (Figure 2.4); a mouse is not just a small rat; still less is a bacterial cell a small human. So we need to look beyond point mutations to understand what sort of changes allowed our common ancestor to have descendants as different from one another as humans and bacteria.

An idea of the problem—and a hint at a possible solution—comes from consideration of what is involved in evolving a new protein with a new function. The new function will nearly always be related in some way to the function of a protein that already exists and is already being fulfilled. For example, hemoglobin and myoglobin are both proteins found in mammals and other vertebrates. We shall consider them later in this book from various points of view, but for the present it is sufficient to say that both have a capacity to bind oxygen, myoglobin rather tightly and without a property known as *cooperativity*, and hemoglobin less tightly but with cooperativity. The details of what cooperativity is can wait until Chapter 10; for the moment it is enough to say that it is important for allowing hemoglobin to do its job well, and would not be a desirable property for myoglobin to have.

The two proteins are quite similar in structure and have similar aminoacid sequences, so it seems almost certain that they are derived from a common ancestral protein that was also able to bind oxygen. Presumably it resembled myoglobin more than hemoglobin, as myoglobin is somewhat simpler in structure than hemoglobin, and it was certainly needed for binding oxygen in the

Fig. 2.4 Dimensions of biological entities. Even when they are drawn to look the same size, animals of the same genus remain easily distinguishable from one another: a domestic cat is not just a scaled-down tiger

ancestral organism. The problem is now how to evolve a new protein, with new properties, such as binding oxygen more weakly but with cooperativity, without losing the old properties that continued to be needed.

Consider a horse breeder who wants to breed a new strain of racehorses from a breed of shire horses, but who still needs shire horses for heavy work on his farm. This is not of course a realistic project for a modern horse breeder, and probably never would have been realistic to expect to achieve within the lifetime of a single breeder, but in past centuries it would have been quite feasible as something to attempt over several generations. Indeed, something of this sort almost certainly did happen before the whole process was accelerated by crossing locally bred horses with imported ones that had desirable characteristics.

In such a case the breeder cannot just breed from the fastest horses in each generation, discarding the others, as this would mean that the still necessary breed of strong horses would be lost. Instead he would have to maintain separate lines of at least two kinds of horses: fast ones for winning races, and strong ones for doing the heavy work. In effect, therefore, the breeder needs to duplicate the stock in an early generation, and then maintain two (or more) separate stocks thereafter. The process of selection is outlined in Figure 2.5.

In the same way a new protein function can only appear without losing an essential existing one if the process begins with a *gene duplication*. If the part of the DNA that coded for the ancestral globin became repeated, so that two separate genes existed to do the same thing, then one of them could be modified in various ways while keeping the other one unchanged; in this way new properties could emerge without loss of the old ones. Alternatively, like a horse breeder who started by separating an original breed of general-purpose horse—reasonably strong and fast, but neither very strong nor very fast—into two arbitrary groups and then trying to breed a very strong strain from one group and a very fast one from the other, one globin line could become more like myoglobin over the same period as the other became more like hemoglobin.

Just as errors creep into the basic DNA-replicating system and result in point mutations, more severe errors also can result in long stretches of DNA being deleted or duplicated. Even entire chromosomes may be deleted or produced in extra copies as a result of a mistake in the replication process. In both cases the result may be disastrous, producing an offspring incapable of living, but occasionally it may be less damaging, and very rarely it may have no harmful effects at all. It is these very rare cases that probably supply the capacity of organisms to acquire new functions in the course of evolution.

We can find evidence for such gross replication errors by comparing the DNA of a primitive organism with that of its more complex relatives. Although there is no necessary relationship between the amount of DNA that an organism has and its complexity, there are examples that fit with naive expectations. One is provided by the lancelet, a small animal found on beaches around the world. This looks superficially like a small fish, but it is not a fish, or even a vertebrate, as it has no backbone. It is a primitive *chordate*, which means that it is an invertebrate member of the phylum that includes the vertebrates as its

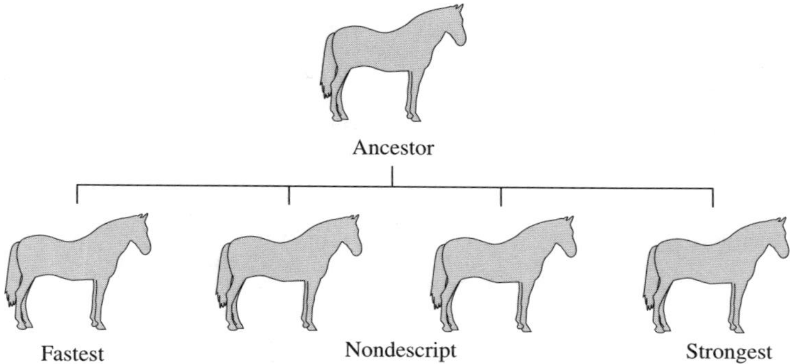

The first generation of progeny will all resemble the parent strain, but one individual will be the fastest, others have no outstanding characteristics, and another will be the strongest

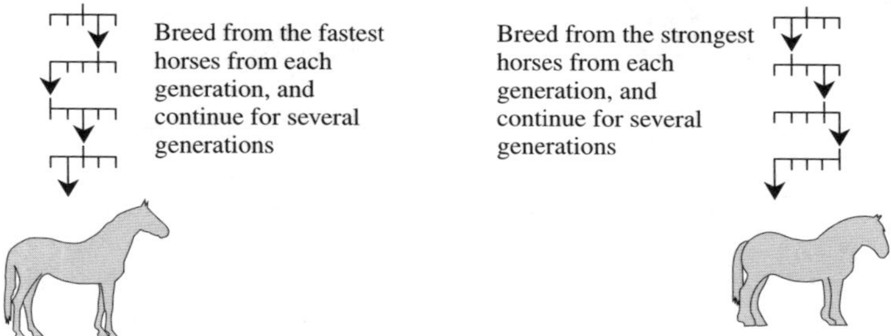

Fig. 2.5 Unnatural selection. The illustration summarizes how a horse breeder can proceed from a single strain of average horses in the first generation to two distinct types in later generations

overwhelmingly preponderant members. It has much less DNA than, say, a human, and within this DNA there is a series of genes known as *PBX, RXR, NOTCH*, and *C*, as illustrated at the top of Figure 2.6. If, now, we look at human chromosome 9, we find that it contains a series of genes called *PBX*3, *RXRa*, *NOTCH*1, and *C*5, similar enough to make it certain that these are the cousins of the similarly named genes of the lancelet. So far this provides no evidence of large-scale duplication, but this comes from examining three other human chromosomes: chromosome 6 contains *PBX*2, *RXRb*, *NOTCH*4, and *C*4; chromosome 19 contains *PBX*4, *NOTCH*3, and *C*3 (*RXR* is missing); and chromosome 1 contains *PBX*1, *RXRg,* and *NOTCH*2 (*C* is missing). More detailed study indicates that chromosome 9 is the one that preserves most of the original structure, but the other three provide clear evidence of ancient replication errors.

As a brief digression I want to comment on the obscure names of these genes: one (*NOTCH*) is vaguely suggestive of something or other; the other three appear completely meaningless. At the beginning such names reflected a lack of knowledge of their functions, but there is an unfortunate habit in modern biochemistry of continuing to use obscure names long after the functions

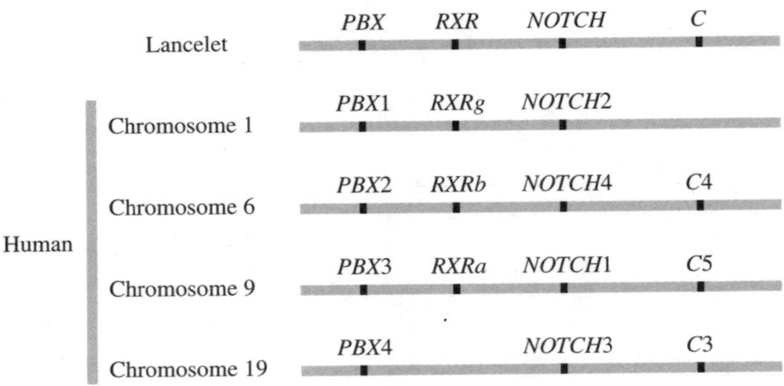

Fig. 2.6 Evidence for gene duplication in human DNA. Four human chromosomes contain three or four each of a series of four genes similar to a series of four genes in the DNA of the lancelet, a primitive chordate

are known. It follows from the conquest of much of biochemistry by molecular biology and the tendency to talk about "gene products" rather than proteins, as if proteins existed only to express what is written in the DNA. This may seem a trivial point (and indeed is considered a trivial point by all those who defend meaningless names on the grounds that "everybody knows" what they refer to), but it is actually a serious barrier to any effort to understand how biological systems are organized. Imagine trying to understand the distribution of telephones in a city in terms of a model that assumes that telephones exist only as devices for giving concrete expression to what is written in the telephone directory.

Returning to the main theme, we can observe large-scale errors in DNA replication happening even today, in humans as well as in animals and plants. Chromosomal abnormalities are actually quite common, and by no means all are lethal. The best known (though not the most frequent) in humans is called *Down syndrome* (in popular accounts it is often called mongolism, an unsatisfactory name that encourages an incorrect perception of the nature of the condition). In its most common form it is also called trisomy-21, and it results from having three instead of the normal two examples of chromosome 21. People with Down syndrome have disabilities, but frequently live into adulthood, so the condition is hardly lethal, and in the last chapter of this book I shall discuss why it should have any biochemical effects at all. Some trisomies involving the sex chromosomes are sufficiently mild to pass unnoticed in some patients. The existence of these and other abnormalities makes it clear that errors involving large stretches of DNA not only occur quite frequently in all species, but also that even if they are nearly always harmful their effects may be so mild that they can pass undetected. In the circumstances, therefore, it requires no great leap of imagination to propose that accidental duplication of genes in the course of evolution has been the motor that has allowed new functions to evolve without loss of existing ones.

3
Adaptation and Accidents

It has been proved, he said, that things could not be otherwise, for everything being made for a purpose, everything must necessarily be made for the best purpose. Notice that noses have been made to support spectacles, and thus we have spectacles. Legs are obviously intended to be trousered, and we have trousers.

Voltaire, *Candide*

First a simple question. Which of these hands are you more likely to pick up?
a. ♠ none, ♥ AKQJ10 98765432, ♦ none, ♣ none;
b. ♠ 92, ♥ Q10943, ♦ 42, ♣ 8762

Zia Mahmood, Guardian bridge column, May 28, 1996

In Voltaire's classic novel the philosopher Dr Pangloss taught that "All is for the best in this best of all possible worlds," a belief he stuck to through every disaster that befell him and his companions, whether expulsion from the ancestral home, dismemberment (more accurately disbuttockment), rape, syphilis, forced labor in a Turkish galley, the great Lisbon earthquake of 1755, or burning over a slow fire (this last intended as a discouragement to further earthquakes).

Dr Pangloss has had many followers in the history of evolutionary biology, who have interpreted every variation seen in nature as an adaptation of some kind. If the left brain controls the right half of the vertebrate body, this must be because that is the best way of arranging matters; if the structure of the hormone insulin is the same in pigs and dogs, but slightly different in humans, it is because humans have different needs from pigs and dogs; if sperm whales have more of the protein myoglobin in their muscles than horses do, it is because they need more of it; if polar bear liver contains so much vitamin A that it is toxic to explorers who eat polar bears, it is an adaptation to protect polar bears from predators; if the aminoacid arginine can be represented in six different ways in the genetic code, whereas the more common aminoacid aspartate can only be represented in two ways, it is because efficient operation of the protein-synthesizing machinery of the cell has more need of redundancy for arginine than for aspartate; if horse liver contains a large amount of the enzyme alcohol dehydrogenase (the same enzyme that makes alcohol in yeast fermentation), it is to allow horses to make as much alcohol as they want without resorting to the methods that humans employ; if 90% of some genes consist of long stretches of bases that appear to code for nothing whatever, the explanation is not that they have no function but that we have so far been too stupid to find it.

I could give many more examples, and I have deliberately mixed them here: there are genuine cases of adaptation (sperm whale myoglobin), sometimes with a wrong explanation attached (horse liver alcohol dehydrogenase), accidents of diet (vitamin A in the polar bear), frozen accidents of evolution (left and right brains, genetic code for arginine and aspartate, and, probably, the structure of insulin), opportunistic or "selfish" behavior of DNA (noncoding sections of genes). Let us look at these in more detail, as otherwise I may appear to be just expressing personal opinions that others are free to disagree with.

Sometimes, as in the case of the sperm whale, we can see a clear and obvious relationship between the lifestyle of the animal and the biological observation. A sperm whale spends a large amount of its life swimming under water, but as an air-breathing mammal it cannot obtain its oxygen like a fish from the water. Swimming under water requires a large amount of oxygen: few people can swim as far as 25 m under water, and even trained swimmers can often do little better than that. But a large whale can swim for as much as an hour without breathing, while expending large amounts of muscular energy. Moreover, we cannot easily imagine a whale adopting a different style of life where it spent most of its time on the surface and breathed more frequently. It has an absolute need, therefore, for a reliable supply of oxygen.

One solution might be to become more like a fish by evolving gills or even a wholly original system for extracting oxygen from water. But evolving gills is not something to be done overnight, or even in the millions of years since the whales returned to the sea: it would require extensive modification of many structures, and is, in short, the difficult solution. An obvious alternative would be to evolve blood with a greater capacity for storing oxygen. It sounds plausible, but it is not as easy as it sounds. Even in humans, with our relatively modest needs for holding our breath, the blood is already so full of red cells, and the red cells so full of hemoglobin, that there is scarcely room for any more. Significantly increasing the oxygen capacity of human blood is not an option, unless we find an alternative to hemoglobin, and it is not an option for whales either.

What the whale has done, therefore, is to increase the oxygen storage capacity of its muscles, which contain a far higher concentration of myoglobin than the muscles of land mammals. Myoglobin is not the same as hemoglobin, but it is quite similar, and is also capable of binding oxygen reversibly. (The two molecules played an important role in the development of understanding how enzymes are regulated, which I shall discuss in Chapter 10.) In land mammals, the primary function of myoglobin appears to be to allow rapid diffusion of oxygen through muscle tissue, but it also helps to increase the total amount of oxygen that can be stored, and, by means of large amounts of myoglobin, whales can store much larger amounts of oxygen than land mammals can manage. Incidentally, although the name of myoglobin is less familiar to most people than that of hemoglobin, everyone has seen it, and its red color, much like that of hemoglobin, is an everyday sight. If you cut a piece of fresh meat, or you start to cook a hamburger, a red liquid flows out: this red liquid is not, as you might easily suppose, blood, but cell water colored by myoglobin. (There may be traces of blood in this liquid, and hence traces of hemoglobin; nonetheless, myoglobin, not hemoglobin, is the main source of the red color.)

If this interpretation of the function of myoglobin is correct, we should expect land mammals to be less crucially dependent on it than diving mammals, and we should also expect small mammals to need it less than large ones do, because the distances that oxygen needs to move across the muscles is much smaller. In this connection, therefore, it is interesting to note that modern genetic techniques have allowed study of mice that lack myoglobin completely, and they turn out to show no detectable differences from ordinary mice, even when exercised, that is to say even in conditions where the need for myoglobin ought to be greatest. However, I shall be surprised if humans who have no myoglobin prove to be healthy, and amazed if whales do.

What about horse liver alcohol dehydrogenase? Drunk wasps are a common sight when the ripe apples fall to the ground in late summer, but who has seen a drunk horse? Horses do not need alcohol dehydrogenase to make alcohol, but to destroy it. It is a detoxification enzyme, and horses need it in large quantities because their bacterial flora produce large amounts of alcohol as an end product of their metabolism (just as it is in yeast fermentation). It is a detoxification enzyme in humans also, notwithstanding the fact that we get most of our alcohol in a different way and we prefer not to think of it as a poison. So it is correct to regard the high level of alcohol dehydrogenase in horse liver as an adaptation, but wrong to explain it in the silly way that I suggested at the beginning. This example illustrates, incidentally, an important point about enzymes that is sometimes forgotten. Like any other catalyst, an enzyme does not determine the direction in which a reaction proceeds, nor even how far it will proceed given enough time. These are questions that are decided by energetic considerations that are independent of whether a catalyst is present or not; the catalyst only determines how fast the reaction proceeds toward equilibrium. Adding a catalyst to a chemical system has very much the same effect as lubricating an engine: it does not suddenly allow you to roll uphill; it just allows you to go more easily in the direction you would take anyway.

However, by no means everything we observe in a living organism is an adaptation, or, if it is, it may be a different kind of adaptation from what it appears to be at first sight. The vitamin A in polar bear liver is not a mechanism to protect the animal from predators: an animal as large as a polar bear has no predators in the normal course of its life (if we ignore occasional unhappy encounters with killer whales), and the unfortunate explorers who fulfilled this role on one occasion did not know about the vitamin A until it was too late; as a result they protected neither the bears nor themselves. In reality, the vitamin A in the bears' livers is just a consequence of eating very large amounts of fish, which, as every child who has been made to consume cod liver oil knows, is an excellent source of vitamin A. In a sense we can still call it an adaptation, but it is an adaptation in the sense of increased tolerance to vitamin A beyond that of other mammals, not an increased need for the vitamin. It seems better to think of it as an accidental consequence of diet.

The existence of sections of noncoding DNA within genes, or *introns*, is hardly an adaptation, at least as long as we think of an organism as being organized primarily for its own good. (Introns are actually just one of various kinds of "junk DNA," which exists in huge amounts and seems to be good for nothing whatever, though we must not, of course, exclude the possibility that at least some of it has a function that has so far eluded detection.) If, on the contrary, we think of an organism primarily as a vehicle to ensure the survival of the DNA that it contains, then it is not so crazy. After all, if a stretch of DNA can ensure its survival just by inserting itself in a gene that is ensuring its survival by doing something useful, why go to all the trouble of specifying a useful protein? Perhaps you will find it difficult to believe that all this DNA is no more than junk, and prefer to think that it has some function that we haven't discovered yet. If so, you may like to reflect on the remarkable fact that the amount of DNA that different organisms contain bears little relation to their complexity. The lungfish, for example, a tropical fish that spends most of its time out of the water, has about 20 times as much DNA in each of its cells as we do ourselves. Which is easier to believe, that a lungfish is 20 times as complex as a human, or that it has 20 times as much junk? Richard Lewontin has doubted whether a dog is more complex than a fish, but, even if we cannot assert with too much confidence that a human is more complex than a lungfish, it is probably not too self-satisfied to claim that we have at least one-twentieth of the complexity of a lungfish. Would anyone seriously argue that the alga *Gonyaulax polyedra*, which has more than six times as much DNA as a lungfish, is more than 100 times more complex than a human?

An accident like the high tolerance of polar bears to vitamin A can fairly easily be reversed. If some descendants of today's polar bears migrate permanently to a different habitat and adopt a different diet, it will not at all be surprising if they lose their tolerance to vitamin A. We can see this kind of effect in the human tolerance to lactose: Europeans, and others like the cattle herders of Africa who consume large amounts of milk after infancy, retain the enzymes necessary for digesting lactose, or milk sugar, into adulthood; in other parts of the world, where milk is regarded as a food only for infants, the capacity to digest lactose is largely absent from adults.

Other accidental changes may have been preserved simply because it did not matter much one way or another. The majority of the differences that exist between the proteins of different organisms, including the insulins of pigs, dogs, and humans, are probably of this kind. Although in the early days of studies of protein evolution, 30 years ago, most people tried to apply the Panglossian pan-adaptationist logic to proteins, nearly everyone has since become converted to the neutralist view that I mentioned in Chapter 2, that most replacements of one protein structure by another have no functional effects one way or the other. This is almost impossible to test, however, because calculations indicate that the advantage needed to ensure that a gene is not replaced by an almost equivalent mutant is so small that very few experiments are sensitive enough to detect it.

Other accidents may result from events so long ago in evolution that altering them now would entail such a complicated series of changes that it is effectively impossible and will never happen. We can easily imagine that a nervous system in which the left and right brains controlled the left and right halves of the body, respectively, might be marginally more efficient than the present arrangement; but what we cannot easily imagine is the process, with all the rewiring that would be needed in many parts of the nervous system, that would allow the change to be made. Presumably the accident arose in a primitive ancestor in which it did not greatly matter how the connections were made: if it had mattered, it ought not to have been too difficult to correct it, but the more complex the descendants of that early ancestor became, the more difficult it became to change the wiring diagram in such a fundamental way. Here we do not know with certainty that there is no adaptive advantage in the wiring system that exists, but we do know that switching to a different one would be difficult or impossible, and we have trouble, moreover, in imagining any adaptive advantage. It seems most reasonable, therefore, to regard such cases as frozen accidents, and refrain from searching for adaptive advantages.

Sometimes, however, the case for a frozen accident appears stronger because we can more easily reconstruct both the mechanism that gave rise to the accident and the reasons why it has become frozen. The genetic code is such a case. It makes little adaptationist sense for arginine to have three times as many codons as aspartate, but Jeffrey Wong has shown that it makes good evolutionary sense nonetheless. His idea is that at an early stage in the origin of life relatively few aminoacids could be separately coded in the primitive DNA. Each of these few, which included arginine and aspartate, could be coded in several different ways: the third letter of the three-letter code was probably little more than a spacer or punctuation mark (as indeed it still is to some extent[1]), and even the first two letters included considerable redundancy.

As life became more complicated, this system became inadequate to the task, because manufacture of more efficient enzymes and other proteins

[1] Out of 16 permutations of the first two bases, 8 fully specify the aminoacid without regard to the third base: for example, TCT, TCC, TCA, and TCG all code for the aminoacid serine, so if the first two bases are TC the third one can be anything. Of the other eight, six result in a two-way selection: for example, CAT and CAC code for histidine, but CAA and CAG code for glutamine. Putting the third base into two classes in this way makes chemical sense, as T and C have similar structures (and are known as *pyrimidine bases*), whereas A and G have a different kind of structure (and are known as *purine bases*). These structures can be seen in Figure 2.1.

required a larger number of different aminoacids to be specified—maybe not all of the 20 that we know today, but more than the primitive organisms needed. How was this greater precision to be achieved without scrapping the whole system and starting again? Wong's idea was that evolution of the genetic code occurred concomitantly with the evolution of new metabolic pathways to synthesize the extra aminoacids that were needed, and that the aminoacids that were coded for in the primitive code would have to "give away" some of their codons to their metabolic descendants. His full hypothesis is more complicated than this, of course, but it explains the observation I started with. On this hypothesis we expect an aminoacid like aspartate, which is the metabolic precursor of several other aminoacids, to have given away so many of its original codons that it now has rather few. On the other hand, arginine is not the metabolic precursor of any other aminoacid, and so we expect it to retain the full stock of its original codons: so even if it had fewer than aspartate to start off with, it has more now. More generally, Wong's hypothesis leads us to expect no particular correlation between the number of codons that an aminoacid has and its frequency in protein structures, but a good inverse correlation between the metabolic usefulness of an aminoacid for making other aminoacids and the number of codons it has, and that is just what we observe.

Why should such an accident become frozen so that now it is impossible to modify? Even the simplest of present-day organisms needs to make a few thousands of different proteins, and more complex ones need far more. The sizes of these proteins vary, but if we take an average protein to consist of 250 aminoacids arranged in a definite order we shall not be grossly far from reality. Of these, perhaps 8%, or 20 per protein, will be aspartates. Suppose now we ask what would be the result of altering the genetic code so that one of the two codons that currently codes for aspartate became ambiguous: instead of coding aspartate all of the time, it started to code for aspartate only half of the time, the other half of the time being read as another aminoacid, glutamate. I have deliberately chosen one of the smallest modifications I could think of: I have not assumed a complete unambiguous switch of one aminoacid for another, and the particular switch is a modest one: glutamate is chemically quite similar to aspartate, but a little larger in size. (Together they are responsible for the basic properties of proteins, but, in a frozen accident in the evolution of biochemical knowledge they are known to most biochemists as "acidic aminoacids"[2].) So we may expect that substituting glutamate for aspartate may have very little effect at many sites, and that the new protein may be just as good as the old.

[2] This is an important misconception that seriously complicates the efforts of students to understand the properties of proteins, but it is not as crazy as it may appear. If the free aminoacids are prepared in their pure forms they are indeed acids—aspartic acid and glutamic acid. However, when they are incorporated into proteins they lose two of their ionizable groups completely, and the third, the one that justifies calling these free forms acids, loses its proton in neutral solution: it then no longer has a proton to donate, so it is no longer an acid; but it can accept one, so it can act as a base.

This, indeed, is what study of protein sequences across different organisms leads us to expect. There are many examples of proteins where the sequence in one organism has an aspartate residue and that in another has glutamate. In fact many observed substitutions are far more radical than that, and we can find quite different aminoacids substituting for one another in perfectly functional proteins. Contrary to what creationists may believe, there is a great deal of redundancy in protein structures, and it is far from the case that a protein sequence has to be exactly right. On the contrary, most protein functions can be fulfilled by a large number of known different sequences, not to mention all of the possible ways that are not yet known.

Let us return to our example of a conservative change to the genetic code that would allow an aspartate codon to be read as a glutamate codon about half of the time. As there are two aspartate codons in the present-day code, and we are assuming one of them to remain unambiguous, this implies that about one-quarter of the aspartates in the proteins of the unmodified organism will be replaced by glutamates in the mutant. For a protein with 20 such residues, this leaves us with about a 0.3% chance that any individual protein molecule will be completely unchanged, and a 99.7% chance that it will contain one or more glutamates where aspartates ought to be. (There is a 0.75 probability that any one site is unchanged, 0.75×0.75 that any two will be unchanged, and so on until a $0.75 \times 0.75 \times 0.75 \times 0.75 \times 0.75 \times 0.75 \times 0.75 \times 0.75 \times 0.75 \times 0.75 \times 0.75 \times 0.75 \times 0.75 \times 0.75 \times 0.75 \times 0.75 \times 0.75 \times 0.75 \times 0.75 \times 0.75 = 0.003$ chance that all 20 will be unchanged.)

How likely is it that an organism could survive this? As I have said, most proteins are not very fussy about which aminoacids occupy most loci, so there is a good chance that some or all of the mutant molecules will be perfectly functional. So, let us guess that there is a 99% chance that the organism will not even notice that a particular protein is modified in 99.7% of its molecules, but a 1% chance that only the completely normal molecule will do. Let us further suppose that for the 1% of proteins that require a perfectly correct sequence to function the 0.3% of such perfect molecules are not enough for the needs of the organism. So, at the end of all this, the chance that an organism can survive is 0.99 raised to the power of the number of different kinds of protein that it needs to make. Even for a bacterium, the number of different proteins is by no means small: for those that have been most studied, such as *Escherichia coli* and *Bacillus subtilis*, it is believed to be around 4000, but free-living organisms exist with only about a third as much genetic material as these, and the parasitic bacterium *Mycoplasma genitalium* has fewer than 500 genes. So let us suppose we are dealing with a free-living organism that requires a minimum of 1300 different kinds of functional proteins in order to live. Then 0.99 to the power 1300 gives us 0.000002, or odds against survival of about half a million to one.

As I have made a number of assumptions in this calculation, I should perhaps pause a moment to examine whether there are any facts that would enable me to estimate whether the calculation is grossly in error. After all, there may be little reason to care if the final result is wrong by a factor of 10, or even 100, but it would be useful to have an idea if it is likely to be wrong by a factor of a million or more. For this we may consider a series of mutants of *E. coli* called *ram* 1. These mutants contain faulty machinery for translating genes into proteins, and although by the standards of most human keyboard operators they are quite accurate, they make far more errors than the normal bacteria make: out of every 100 molecules that they make of a 1100-aminoacid protein called beta-galactosidase (much larger than what I have taken as average), no two are identical. These bacteria grow poorly, and in a natural environment would rapidly be overwhelmed in numbers by their healthier cousins, but they do grow, and in the artificial conditions of the laboratory they can survive. We are far from the creation scientists' fantasy of proteins that have to be exactly correct in every detail for the organism to live and reproduce. This example is from Jacques Ninio's book *Molecular Approaches to Evolution*, and he goes on to discuss a more technical example that is even more relevant to the problem of surviving a change in the code. This involves the use of molecules called *missense suppressor tRNAs*, which have the effect of causing certain codons to be misread with a frequency of around 10%. Bacteria can survive the presence of one such suppressor tRNA, but not two. Again, this implies that although an appreciable error rate can be tolerated, there are limits that cannot be exceeded.

In general, we can conclude that for obtaining a rough idea of the probability that an organism could survive a change in the genetic code, the earlier calculation is not unreasonable. Thus even a simple organism has a low probability of being able to tolerate a minimal change in the genetic code, and for a complex organism the chance would be much smaller. Nonetheless, it is not so small as to be absolutely negligible, and so we might expect to see one or two vestiges of alternative codes in the simplest of organisms.

In fact, as mentioned in Chapter 2, the so-called "universal" code is now known to be not quite universal. In bacteria, and in the nuclei of all the cells of all the nucleated organisms that have ever been studied—and here I do not just mean everyday animals like cats, dogs, cows, horses, etc., but organisms as diverse as butterflies, cabbages, eagles, mushrooms, otters, crabs, sea urchins, cucumbers, everything—the genetic code is exactly the same in all its details. However, not all genes are in the nucleus. As mentioned in Chapter 2, a few proteins are coded in the small bodies within cells known as mitochondria, and in these the code is very slightly different from the "universal" code. Mitochondria are believed to be remnants of formerly free-living organisms, probably primitive bacteria, that have become so fully adapted to the habit of living within cells of other organisms that they are now part of them. Whether their slightly different codes result from experimenting with alternative codes, or they are so ancient that they came into existence before the "universal" code became frozen is a question I shall return to shortly, but it does not matter for the immediate point: they confirm that even though the chance of surviving a change of code may be small, it is not zero.

If Wong's hypothesis is right, the genetic code is not absolutely immutable, and it should be possible to modify it experimentally. He has shown that this can be done, albeit with a much more modest change than the introduction of ambiguity into an aspartate codon that I considered earlier. Tryptophan has the most elaborate structure of the 20 aminoacids needed for proteins, and is one of those that is used least frequently. It has only one codon in the universal code, and was almost certainly one of the last aminoacids to be recognized in the code. The aminoacid 4-fluorotryptophan is chemically almost the same as tryptophan, but it has one of the hydrogen atoms of tryptophan replaced by a fluorine atom. Because of this high degree of similarity, we may expect that in nearly all cases a protein with some or all of the tryptophans replaced with 4-fluorotryptophan would be functionally indistinguishable from the native protein. Nonetheless, in normal bacteria such as *Bacillus subtilis* the enzyme that starts the incorporation of tryptophan into growing proteins does not readily recognize 4-fluorotryptophan, and the fluorinated aminoacid is only very rarely incorporated into protein even if the bacteria are grown in a medium rich in 4-fluorotryptophan but lacking tryptophan. Such a medium does not exist naturally, but it can readily be produced artificially in the laboratory, and we might expect that bacteria growing in it would find it advantageous to recognize 4-fluorotryptophan, thereby avoiding the need for the metabolically expensive synthesis of the natural aminoacid. Wong found that this was indeed the case, and in careful selection experiments he was able to go from a natural strain of *Bacillus subtilis* with a 700-fold preference for tryptophan to a mutant with a 30-fold preference for 4-fluorotryptophan.

Is there any evidence that the same sort of evolution may have occurred under natural conditions? The rare aminoacid selenocysteine may provide an answer. This is the same as the "standard" aminoacid cysteine except that it has a selenium atom in place of the usual sulfur atom. Most organisms do not incorporate it into their proteins, and their genetic codes do not recognize it. However, the few organisms that use it do incorporate it into proteins and they do so by the normal genetic machinery. (That is to say, they do not use special mechanisms to alter the protein after it has been made, in contrast to certain other unusual aminoacids, which are *not* coded in the DNA: an example is hydroxyproline, an essential component of the fibrous protein that gives a muscle tendon its tough stringy character, which is called collagen.) In fact, such organisms have transferred to selenocysteine one of the two codons normally used for cysteine. We can imagine that the original switch may have been brought about by natural selection pressure similar to the artificial pressure used by Wong to change the codon for tryptophan into a codon for 4-fluorotryptophan.

Before continuing, I want to look at the apparently trivial bridge problem that I quoted at the beginning of the chapter: which is more likely, that you will receive a hand consisting of all 13 hearts, or the second hand, which lacks any very obvious feature? The naive answer is that the nondescript hand is more likely, because it is an "ordinary" hand that no one but a bridge expert could remember for 5 min, let alone the next day, whereas if you picked up all

13 hearts you would be telling your friends about it for the rest of your life. A more sophisticated answer is that as long as the cards were properly shuffled before dealing them the two hands are exactly equally likely and you should receive each of them once on average in every 635 013 559 600 deals; the fact that we perceive one as more improbable than the other just reflects the fact that there are vastly more of the boring hands than there are of the interesting ones.

In reality the second answer is also naive, and is also wrong. The correct answer is that you are more likely to receive the 13 hearts, as is borne out by the fact that several hands of this kind are reported to have been dealt in the history of bridge, even though it is quite unlikely that 635 013 559 600 hands of bridge have yet been dealt. As Zia Mahmood commented in the original bridge column, "there is a small chance that someone will have stacked the deck so that you get all thirteen hearts, while nobody is going to arrange for you to be dealt hand *b*!" The point is that for calculating the odds I slipped in a qualification "as long as the cards were properly shuffled before dealing them" that I hoped might sound so reasonable that it would pass unnoticed. In fact this sort of qualification is *not* reasonable, either in games of bridge or in the evolution of life, because there are always some biases that cause apparently equivalent outcomes in a random test to be unequally likely. You should always be suspicious of statistical calculations and search for possible biases that can cause one-in-a-million events to occur much more often than once in every million trials. This is not, I hasten to add, intended as an argument against making estimates of probabilities—far better base your decisions on some sort of calculation of chances than on a hunch or the pattern of tea leaves in a cup—you just need to avoid placing too much trust in the calculations. I am writing these words a few days after the rocket Ariane V exploded, destroying some $2 billion worth of investment in about a second: clearly someone put too much confidence in their risk calculations!

When I was calculating above how likely it was that a bacterium could survive a change in code, I assumed that the two codons for aspartate were equally used, so that a change in the code would affect about half of all the aspartates in the proteins of the organism that suffered the change of code. In reality, however, codons that code for the same aminoacid are by no means equally used: there is no doubt that codons are unequally used, and although the reasons for this have not been thoroughly worked out there is evidence that organisms prefer certain codons for proteins that are being made in large quantities, but different ones for proteins being made in small amounts.

It is quite possible, then, that variations in codon usage might cause a particular codon not to be used at all in a particular organism. This is not purely hypothetical: for example, the mitochondria of the yeast *Torulopsis* appear to avoid using four codons entirely. Why this should happen in the first place is unclear, but it might, for example, permit some simplification of the recognition machinery. Once it has happened, it has one immediate consequence: a subsequent evolutionary event causing the unused codons to be used again for any aminoacid at all will have no effect on the correct coding of any existing proteins, and thus will have no deleterious effect on the viability of the organism. The "universal" genetic code begins to look rather less frozen than it once did, and in fact numerous variations have been found in mitochondria.

There are two reasons why we should not be surprised to find these variations among mitochondria rather than elsewhere. First, the number of different proteins coded by mitochondrial DNA is very small, far fewer than the 4000 assumed previously, and as the calculation showed, the risk incurred by changing the code increases steeply as the number of proteins concerned increases. (To avoid implying something that is not in fact true, I repeat a point made briefly in Chapter 2, that mitochondria get many of their proteins from the host cell, and these proteins are coded in the nucleus with the usual "universal" code. The proteins that are coded by the mitochondrial DNA constitute a small fraction of the proteins that work in the mitochondria.) Second, the small number of mitochondrial genes makes it much more likely than it would be for the nuclear DNA that certain codons should fall into disuse as a simple statistical fluctuation, thereby facilitating the sort of mechanism I have just discussed. These sorts of considerations suggest that it is quite possible that the mitochondrial variations arose after the universal code became established, though they do not, of course, exclude the alternative supposition that they are survivals of extremely ancient versions of the code.

I have spent most of this chapter discussing examples of accidents rather than of adaptation, and without even mentioning the main theme of this book, in order to emphasize at the outset that this is not a Panglossian book. I am not going to find optimality everywhere I look, and I am not going to suggest optimality without providing evidence of how it is assessed. In other words, I am going to describe examples where we can show beyond any doubt that although other metabolic designs might be possible, and might work adequately or even very well, the designs that successful organisms actually use are the best possible ones. As I have already discussed in this chapter that accidental choices can very plausibly be frozen in evolution, I have to show how in many cases organisms have nonetheless sought out the best solution, even though replacing a suboptimal design that actually worked quite well may have required complicated adjustments. Finally, by studying where in the living world we may expect to find exceptions to the general rules, organisms that survive while retaining suboptimal solutions to biochemical problems, I believe that we can shed light on some unsolved questions of evolution.

Nonetheless, I ask you to approach what I have to say with skepticism, to look for the guiding hand of Dr Pangloss in forming my opinions, to ask whether the facts I shall describe can be explained without invoking optimality principles. To encourage such a healthy skepticism, I shall devote Chapter 7 to examining entropy–enthalpy compensation, a quite false example of optimization that beguiled many distinguished biochemists in the 1970s (and still retains a few today). I hope I shall convince you that my examples of optimization can survive all your skepticism.

4
Metabolism and Cells

Around 20 years ago, Departments of Biochemistry around the world started to rename themselves Departments of Biochemistry and Molecular Biology, a reflection of the changing status of classical biochemistry, with its roots in physiological chemistry, as compared with the growth of gene-related biochemistry. In the days when biochemists were not ashamed to call themselves biochemists, you could go into any biochemistry laboratory in the world and find a Chart of Metabolic Pathways on the wall. You can still find these old relics, the paper brown and curling, in forgotten corners of today's biotechnology institutes, but they are liable to have "Copyright 1972" printed on them, and to have avoided being replaced by maps of the human genome only by luck.

This is a pity, because even the most abbreviated charts of metabolic pathways (and even the largest are to some degree abbreviated) impress you immediately with the sheer number of different chemical reactions that are in progress in any living organism, many of them going on simultaneously in the same cell. In an organism like a bacterium, in which one cell constitutes the whole organism, one cell has essentially the same capabilities as any other; but in organisms such as ourselves, in which the individual is a collection of many cells, the cells are by no means all equivalent. Some are more specialized than others—some deal with movement, others transmit stimuli, store fats, regulate the temperature, emit light, or accumulate electric charge, etc.—but all are in constant chemical activity: they move, control what passes through their membranes, degrade substances to produce energy, which they use to accumulate reserve materials, for the continuous synthesis of their own materials, for their own internal contractile activity, and for many other functions.

A very simplified example is shown in Figure 4.1, in which the different substances involved are just represented by gray balls, with lines between them representing chemical reactions; in a real chart for teaching metabolism the names and other details would be shown explicitly. This diagram may look very complicated, and you may suspect that I have chosen an unnecessarily complicated example, but in fact the reverse is true: the organism represented, the form of the parasite *Trypanosoma brucei* that exists in the bloodstream of people infected with African sleeping sickness, has possibly the simplest

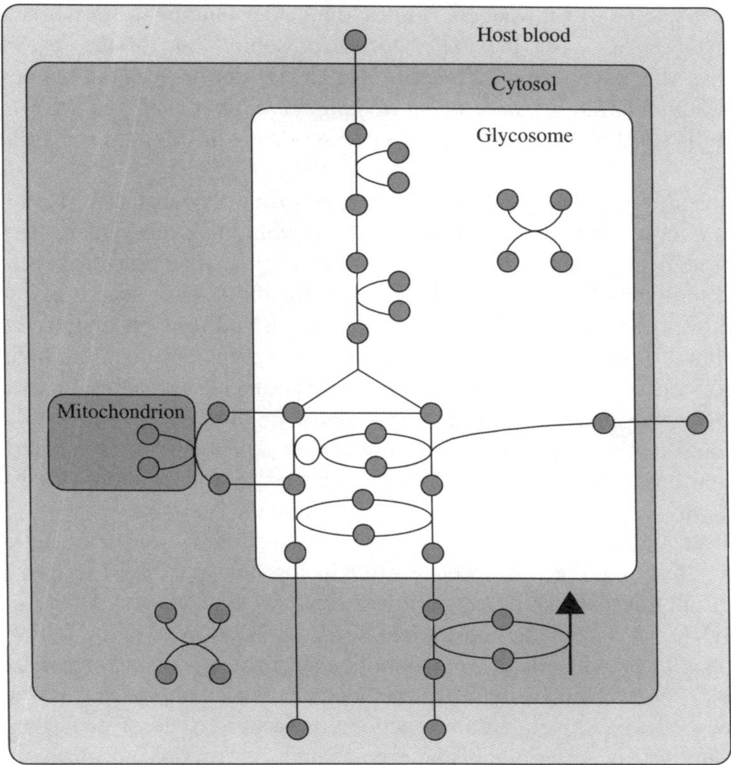

Fig. 4.1 Simplified representation of the main metabolic activity of the form of the parasite *Trypanosoma brucei* found in the bloodstream. The different degrees of shading distinguish the four different compartments in the system, including the blood of the host, which supplies the glucose and inorganic phosphate needed by the parasite for its metabolism, and receives the pyruvate and glycerol that it excretes. The shaded circles represent the different metabolites, connected by lines to indicate enzyme-catalyzed reactions

metabolism known, and only its major chemical activity is included. In the bloodstream it depends on its unfortunate host for most of what it needs, and it excretes a substance known as pyruvate that most organisms treat as being much too valuable to throw away. Yet even in this grossly oversimplified example you can see that numerous reactions are in progress, and that they do not even happen in the same place, but in three different compartments of the cell, together with the environment provided by the host.

All of this chemical activity—normally far more complicated than what I have shown—constitutes the machinery known as *metabolism*. Its reactions are highly specific, as each enzyme is itself highly selective for the substrate of its reaction. Inside a cell the many chemical reactions proceed at once, at similar rates, without interference or obstruction. The organization of all this chemistry is, in effect, what constitutes life. Understanding the principles of its organization has been the major focus for some years of the research of Enrique Meléndez-Hevia in Tenerife. He has set out his ideas in his book *La Evolución del Metabolismo: hacia la Simplicidad*, but as this is not available in English I shall deal with them in some detail in this and the following chapter.

According to the great geneticist Theodosius Dobzhansky, "Nothing in biology makes sense except in the light of evolution"; this remains the central biological truth that has to be understood by anyone who seeks to rationalize biology. (This sentence is often quoted: if you have seen it before you may have recognized it when it appeared in garbled form in Chapter 1.) The question of how life has evolved is open to scientific investigation, both with experiments and by mathematical analysis. Experiments may show, for example, whether environmental conditions postulated to have existed at the origin of life could really have resulted in spontaneous appearance of aminoacids and nitrogen bases. In some cases mathematical analysis can show whether the actual way in which metabolism is organized is the best possible way.

Given the apparent completeness and perfection of modern metabolism, we may fear that there is no hope of deducing how it evolved to its present state, if all traces of its more primitive states have disappeared. However, just as mitochondrial genetic codes yield some clues as to how the "universal" code may have evolved, so we may hope that some organisms preserve some traces of more primitive types of metabolism. For example, an enzyme that was needed to catalyze certain reactions in earlier stages of evolution may have survived in some organisms even though its reaction no longer fulfills any function and has been eliminated from most organisms.

As I have emphasized, the metabolic chart illustrated earlier in this chapter is unusually simplified. A more realistic one would not only include names of the molecules transformed in the reactions, but it would probably include their chemical structures as well, and would identify the enzymes that catalyze the different reactions. More important unless it was clearly intended to represent just a small part of the metabolism of a cell, it would include many more reactions than the 20 or so that I illustrated. Unfortunately, however, it is difficult to put much more on a small page without making the type too small to read, so you will have to imagine how a chart occupying an area of the order of 1 m^2 might look on the wall of a laboratory. The designer will not have shown all of the chemical reactions as having equal status (any more than a modern road atlas shows highways and farm tracks with equal prominence) and will not have scattered them arbitrarily over the paper. This is especially true if you look at a chart designed by someone like Donald Nicholson who has spent many years thinking about how to convey the central ideas in the clearest possible way. (His charts were published for years by the Koch–Light Chemical Company, but the copyright is now owned by the International Union of Biochemistry and Molecular Biology, which makes them available on the web, at http://www.tcd.ie/Biochemistry/IUBMB-Nicholson/). Instead certain groups of reactions will be given higher status by being placed more centrally or printed in larger or heavier type, whereas others will be made to appear more peripheral.

To some extent this may be subjective, imposing a human interpretation on the observations, but it also reflects an objective reality: some reactions carry a greater flux of metabolites than others, some are active in a wider range of cell types than others, and so on. To understand the entire chart, therefore, it is useful to collect the reactions into groups of transformation sequences known as *metabolic pathways*. The number of steps considered to be one pathway can be very small if very few steps are needed to convert one important metabolite into another. For example, serine biosynthesis is a three-step pathway, in which the aminoacid serine is synthesized from 3-phosphoglycerate. At the other extreme, beta-oxidation, the process that converts fatty acids from the form in which they are stored in fat cells into the form in which they are metabolically active, involves seven repetitions of the same four types of step, making an unbranched pathway of nearly 30 reactions.

The pentose phosphate cycle is the pathway that Meléndez-Hevia and his colleagues have analyzed most thoroughly in their efforts to understand the design of metabolism. It consists of 11 successive reactions, in two distinct phases: in the first (oxidative) phase, glucose (a *hexose*, or sugar with six carbon atoms in each molecule) is converted into ribulose (a *pentose*, or sugar with five carbon atoms in each molecule); in the second (nonoxidative) phase, the carbon atoms of six pentose molecules are rearranged to produce five hexose molecules, which allow the cycle to begin again. Figure 4.2, which represents this pathway, doubtlessly appears complicated and unmemorable, and that is no illusion: it *is* complicated and unmemorable. For the student trying to learn metabolism it provides the archetype of an arbitrary, meaningless, and unmemorable collection of reactions, but it will appear much less arbitrary and meaningless (though still perhaps unmemorable) if we take the trouble to analyze it.

The pentose phosphate cycle has various functions, which are different in different types of cells. One of these is its coupling with the synthesis of fatty acids, which constitutes a good example of the harmony in coupling of metabolic pathways, and of the specialization of cells for it. The oxidative phase of the cycle, converting glucose into ribulose, produces "reducing equivalents" of a molecule commonly known by its initials as reduced NADP, which is needed for the synthesis of fatty acids; it is a very necessary coupling, given that the pentose phosphate cycle is one of the main sources of reduced NADP in the cytoplasm of cells where fatty acids are synthesized. The mammary gland, adipose tissue and the liver are the organs in which the synthesis of fatty acids is most highly developed, and they are also the organs with the highest activity of the pentose phosphate cycle.

The red blood cells are the simplest cells in the mammalian body; they have lost the nucleus, and as they cannot renew their material their lifetime is short, only 4 months. They do not synthesize fatty acids and have no respiratory activity. The energy that they need—primarily to maintain their membranes in

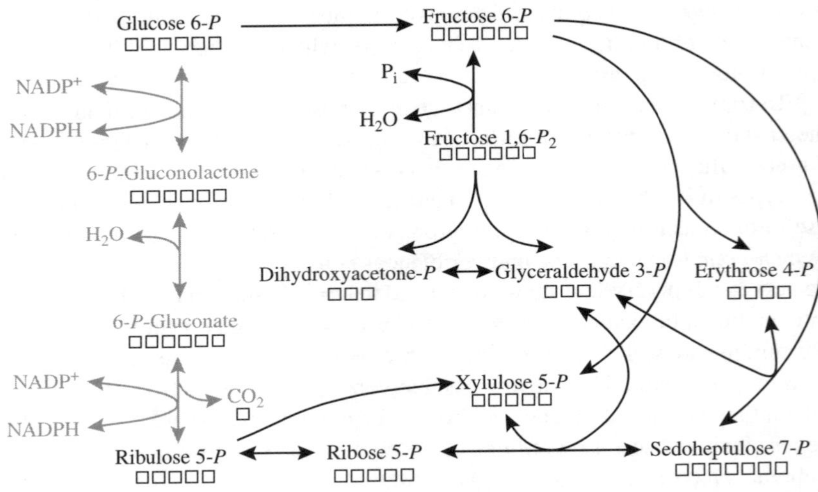

Fig. 4.2 The pentose phosphate pathway. The squares under each chemical name indicate the number of carbon atoms that each molecule contains: glucose 6-phosphate, for example, is a hexose, with six carbon atoms, ribose 5-phosphate is a pentose, with five, and so on. The names shown in gray at the left of the diagram belong to molecules in the oxidative part of the pathway, which is not considered in the analysis in the text. The remainder, the nonoxidative part, can be interpreted as a scheme to allow the hexoses and pentoses to be converted into one another

their correct electrical state and for transporting substances across them—they obtain by converting glucose into lactate without consuming oxygen, a process known as *anaerobic glycolysis:* "anaerobic" means without air, but, as oxygen is the component of air that is relevant here, in practice it means without oxygen; "glycolysis" means mobilization of glucose. The energetic yield from anaerobic glycolysis is feeble, but as the red cells are in the bloodstream itself they have no problem with fuel supply. It may seem paradoxical that the cells that have the most oxygen available to them make no use of it in their own metabolism, but this can be explained in terms of the total specialization of red cells for transporting hemoglobin: the fewer other proteins and other components of all kinds they contain, the more space they have for packing in the maximum quantity of hemoglobin.

Certain genetic deficiencies that affect the pentose phosphate cycle produce severe anemias, thereby demonstrating that the pentose phosphate cycle is essential in red cells, but what is it for, if there is no synthesis of fatty acids? Reduced NADP is needed for other things in addition to this: in the red cell it is needed for repairing oxidation damage to hemoglobin. Although hemoglobin is a protein, it contains a nonprotein part called *heme*, and it is to the heme that oxygen is attached. In the center of each heme is an iron atom, which is liable to destruction by oxygen in a process akin to rusting, and reduced NADP is necessary to repair this damage. It is no coincidence that when blood is dried in the air it turns a rusty color: it owes this color to the same type of oxidized iron that gives rust its color.

The pentose phosphate cycle has other functions in other cells, such as production of precursors for synthesis of various compounds in plants and bacteria, or to feed the photosynthetic cycle in green plants, but to understand how it came into existence in evolution we need to know what its original function was. The answer comes from recognizing that in nature there are just two kinds of sugars that exist in significant quantities, the pentoses, with five carbon atoms in each molecule, and the hexoses, with six. Other sugars exist, with three, four, or seven carbon atoms per molecule, but these are just transient metabolic intermediates, present in tiny amounts compared with the hexoses and pentoses. These two major kinds of sugars have clearly different functions. The major pentoses, ribose and deoxyribose, are components of nucleic acids and thus constitute part of the information system of life; the hexoses are used as fuels, both free as glucose for immediate use and accumulated into polymers like glycogen or starch for long-term energy storage. In addition, they are constituents of quite different polymers whose role is structural: cellulose, the principal building material in plants, is a polymer of glucose; chitin, used for making the shells of crustaceans, mollusks, and insects, is a somewhat more elaborate polymer of molecules closely similar to glucose.

These two major classes of sugars, hexoses, and pentoses, would be totally separate in nature, forming two pools of organic carbon, if there were no metabolic pathway capable of interconverting them. Although we can conceive of a living system organized in this way, it would represent substantial duplication of effort, and would make nutrition a more complicated science than it is. Instead of just labeling processed food with their carbohydrate content, as at present, we should need separate labels for hexoses and pentoses, to ensure that the two kinds of carbohydrates were properly balanced. However, the two classes are similar enough for each to represent a potential reservoir of the other if needed, and it is precisely the function of the pentose phosphate cycle to provide a route between them. This interface between the two great classes of sugars was without doubt its original and primary function.

Evolution and the development of life requires continuous adjustment of the quantity of carbon dedicated to information and the quantity of carbon dedicated to energy and structure. Photosynthesis, the process whereby green plants use light energy to transform carbon dioxide in the air into organic carbon derivatives, involves sugars of various sizes—trioses, pentoses, and hexoses—but the end result, the storage carbohydrate in plants, is starch, a form of hexose, and when sugars are broken down it is always in the form of hexoses. The pentose phosphate cycle, functioning in all organisms, continuously regulates the quantity of glucose that has to be converted into ribose for synthesizing nucleic acids, or the reverse pathway, when nucleic acids are being degraded. This need must have existed very early in the development of life, because management of energy and information is fundamental in any kind of life that we can imagine.

Some primitive organisms—sea urchins for example—consist of a reproductive system, a protective shell, and not much else; others, such as ourselves, lead much more complex lives, with so many other components that the reproductive organs represent a very small proportion of the total mass. Nonetheless, we can eat sea urchins, and other pentose-rich foods like fish roe and yeast extract (marmite), without burdening our systems with excess pentoses, because we can convert the pentoses to hexoses.

Put at its most simple, the reactions of the pentose phosphate pathway can be considered as a series of shuffling steps in which units are exchanged between five- and six-unit structures. In this form the pathway lends itself readily to mathematical analysis, because we can ask whether the particular set of exchanges is the most efficient that can be conceived. This will be the subject of Chapter 5.

5
The Games Cells Play

The road to simplicity has not been easy. In reality, it has been a twisted path, full of obstacles and mistakes, including wrong turnings, bad solutions that were only revealed to be bad when better ones were found.
Enrique Meléndez-Hevia, *Evolución del Metabolismo*, 1993

Analysis of the pentose phosphate pathway needs techniques similar to those used for analyzing mathematical games. Such games have been popular since the nineteenth century, and are still so today. Many have appeared in the pages of such magazines as *Scientific American*, especially during the long period when its *Mathematical Recreations* section was written by Martin Gardner. One class of mathematical game is highly pertinent to metabolism, and involves minimizing the number of steps needed to transform one arrangement of objects into another. The archetype of a combinatorial game is the famous problem of the farmer who must cross a river in a small boat with a fox, a goose, and a sack of oats.

The fox cannot be left alone with the goose, the goose cannot be left alone with the oats, but the boat is too small to carry more than the farmer and one of his three burdens at once. How can he get everything across the river without mishap? The problem is very simple to solve, and most of us have solved it already as children. Nonetheless, it is worthwhile analyzing it in detail, as this will illustrate some points useful for more difficult problems. For the first river crossing, there is only one possible move, for the farmer to cross with the goose. Anything else will allow the fox to eat the goose (if the farmer takes the oats), or the goose to eat the oats (if the farmer takes the fox), or the possibility of either (if the farmer goes alone). The second step is almost equally clear: although the farmer could bring the goose back with him, this would just recreate the original state and thus cannot be a step toward solving the problem. So the second step must be for the farmer to return alone. At this stage, for the one and only time in the game, there are two different but equally good choices, to cross with the fox or to cross with the oats. Whichever he does, he is faced with only one possibility for the next step, which is to take the goose back with him, as otherwise the fox will eat the goose or the goose will eat the oats. In the next step, to return with the goose restores the previous state, so it is not useful, and so he must take whichever out of the fox or oats is left. The two pathways inevitably come together at this point to yield the same state, in which the goose is the only item left on the original side of the river. The ending is then simple: the farmer returns to this side and collects the goose.

There are many variants of this problem. We can make it more symmetrical (as well as making it a more realistic problem for a real farmer to want to solve) by replacing the fox by a second sack of oats or by a second goose. All of them have essentially the same solution, all of them involving more river crossings that appear at first sight to be necessary. The last variant, with two geese and one sack of oats, and thus a single resource that needs to be protected from multiple dangers, makes a convenient introduction to the problem to be considered next. Suppose that I want to send you a chest containing some valuable jewels. I am unwilling to entrust it to the mail, for fear of theft. I have a padlock of high quality that I am confident is impossible to open without the use of a key that is in my possession. However, that seems to be of little use, because if I lock the chest you will not be able to open it unless I send the key as well, but then I run the same risk as before, that the key may be stolen. So the existence of a secure padlock does not seem to solve the problem, and if you have a similarly secure padlock that only you can open, that will not solve it either, as I will not be able to put the jewels into a chest that I cannot open. How can I send you the chest without risking that it can be opened in transit? This does not have the same logical structure as the problem of the farmer and the river, but it is similar enough that you should not have too much difficulty in deducing the answer—which I will come to shortly.

Remarkably, however, a problem that is logically equivalent to the problem of the jewels was believed to be insoluble for 2000 years. When I was a child in the 1950s the problem of the farmer crossing the river was regarded as simple enough for children's magazines, but the logically similar problem was not solved at all until the middle 1970s. In the form in which it is usually posed it concerns methods for sending and receiving secure messages, that is, messages that cannot be read by anyone other than the intended recipient, without requiring either sender or recipient to possess the key needed for deciphering messages enciphered by the other. Exchanging keys is very expensive for governments, financial institutions, and so on that send large numbers of secret messages, but, in the words of Simon Singh in *The Code Book*, "it seems that the distribution of keys is unavoidable. For 20 centuries this was considered to be an axiom of cryptography—an indisputable truth." Even if we discount the first 19 of these centuries on the grounds that before the work of Georges Painvin in France during the First World War the modern scientific study of cryptography had hardly begun, it remains true that for at least 60 years the greatest minds in cryptography believed it impossible.

Yet with the example of the farmer and the river in mind, the solution to the problem of the jewels seems quite easy, as illustrated in Figure 5.1: I send

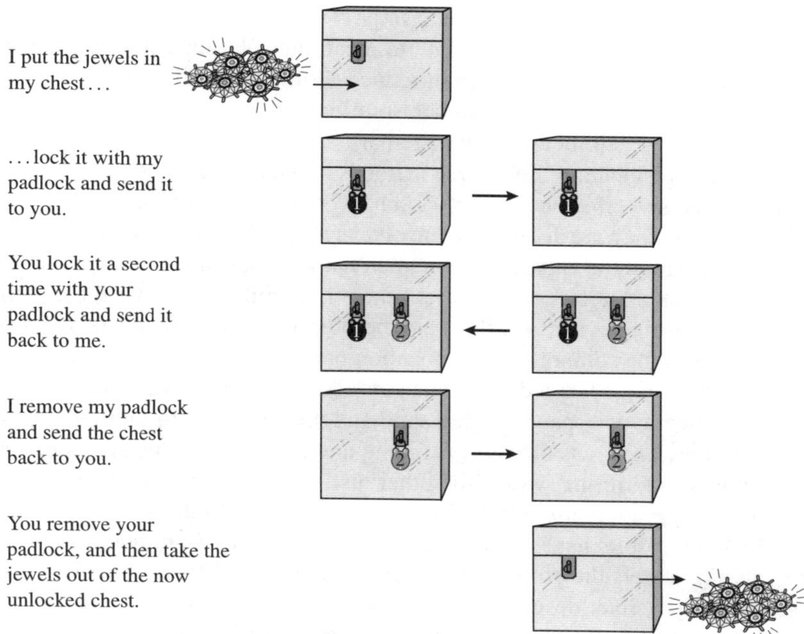

Fig. 5.1 Sending locked goods without sending the key. Use of two different locks allows goods to be sent in a locked container without any key leaving the possession of its owner

you the jewels in a chest locked with my padlock, and you send it back to me unopened but now locked with your padlock as well as mine. I remove mine and send the chest back to you, which you open without difficulty after removing your padlock. This was the problem that was solved by Whitfield Diffie, Martin Hellman, and Ralph Merkle in the 1970s as the first (and in retrospect easiest, though it did not seem like that at the time) step in the development of public key ciphers—now the standard method of sending secret information without risk of having it stolen.

Problems of this kind illustrate several points that are worth being explicit about. First, the solutions involve steps that appeared to be in the "wrong" direction, bringing the goose back to where it started, for example: having to realize this is, of course, the whole point of the original problem, as everything else is obvious. Second, although the solution can be found simply by thinking logically, it can also be found by brute force, just listing all the possibilities at every stage and not thinking at all. Inelegant though it is, this is often the best method for dealing with complicated problems, and the only one that is easy to implement in a computer. Third, problems of this kind always have an infinite number of solutions, as you can always return to earlier states by reversing the previous steps, or by going around a cycle in a branched part of the solution. Consequently it is often not sufficient to ask for a solution: we ought to find the best solution, which implies, of course, that "best" must be defined. In general I shall take "best" to mean the solution with the smallest number of steps, but I may need to introduce additional criteria in cases where this definition produces more than one solution.

It may also be helpful to amplify a remark in the preceding paragraph, about the ease of implementing brute-force methods in the computer. We are often so impressed by how effective computers are at certain kinds of tasks that we may forget how bad they are at ones that involve any real thinking, even thinking at a level well within the capacity of a human baby. Understanding why computers are good at some tasks and bad at others is useful for its own sake, but it has an additional relevance for this book because it helps us to understand how evolution finds the solutions to optimization problems.

Many dogs have learned to do such tasks as make their way along a busy street to collect a newspaper from a shop and return with it to their homes. If necessary they can cross the road, avoiding any cars that may be approaching, and without interfering with any other users of the road that they may encounter. No computer-controlled robot is yet able to do this—at least, none for which the results have been made publicly available. I am not, of course, privy to any secrets the armed forces may have, but even there some speculation may be possible: despite all the boasting that went on at the time of the Gulf War and in more recent military actions, there is little to suggest that anything resembling a weapon with the intelligence of a trained dog yet exists. Chess-playing programs have advanced enormously over the past two decades, but a grandmaster still has a good chance of beating even the best of such programs. Moreover, a program that can play chess cannot normally do anything else, whereas a human chess-player can do a great many other things, such as going to a shop to buy a newspaper, for example.

Despite great efforts to move away from a brute-force approach to computing, most programs still rely heavily on it. What does this mean? It means that successful analysis of a problem involves examining a huge number of possible solutions, testing them according to some preset criteria, and finally selecting the solution that gives the highest score according to these criteria. Unlike a human, a computer program can do all the necessary testing accurately, does not forget which possibilities have already been tried or which solutions have yielded the best scores, and does not get bored or frustrated. Moreover, it can do all this at enormous speed, so fast that the observer can be fooled into thinking that some thought or intuition has gone into the result. But in reality none of this requires any thought at all, and until we can take it for granted that robots can do the sort of tasks that we take it for granted that dogs can do there will be no reason to attribute any intelligence to them.

Natural selection works in a similar sort of way. It differs from a modern computer in two respects, however. First, it is extremely slow, but that is of no great importance as enormous amounts of time have been available—some 4.5 billion years since the formation of the earth, and perhaps half that time since the first living organisms appeared. Second, natural selection has no memory, and thus no capacity for knowing which possible solutions have been tried and found wanting in the past. Instead, it tinkers, changing things[1] here and there in a haphazard way. If, as is usually the case, the modified organism is less capable of surviving than the original, it is discarded, but when, rarely, it is just as good or better, it has some chance of leaving more descendants than the original and eventually replacing it.

Exactly this type of behavior is seen in the response of bacteria such as the pathogen *Salmonella typhimurium* (the cause of some kinds of food poisoning) to chemicals in the mixture of water and nutrients in which they grow. When a colony of bacteria grows in a mixture containing a gradient of glucose, which they consume as food, the cells move, apparently steadily, in the direction of the glucose, a property known as *chemotaxis*. However, this picture is very misleading, as you can see by watching a single bacterium under a microscope. It can swim in a straight line, using a whip-like tail known as a flagellum, but it has no idea which is the right direction to find more glucose, and even if it has once found a "good" direction (increasing glucose concentration) it does not remember it afterwards. After each period of swimming it "tumbles," or rotates at a single point, and then sets off again in a completely random direction—the new direction is not related either to the previous direction or to the direction where the highest glucose concentration can be found; it is just arbitrary. How then is it possible for the population as a whole to move in the right direction? In fact, each individual bacterium does move on average in the right direction, albeit in a very drunken way: it does so because it is able to recognize whether its present state is better (more glucose) or worse (less glucose) than it was a few seconds ago. If things seem to be getting better it swims for a longish period, whereas if they are getting worse it stops quickly and tumbles. This kind of movement is called a "biased random walk": for an intelligent organism it may seem a clumsy way of reaching a goal, but it works, and that is all that matters. It is a little like the procedure used by people who have lost their way in a wood, trying different directions at random in the hope of judging whether they are getting closer to or further away from where they want to go. However, it works better for a simple organism than for an intelligent one, as intelligent organisms are liable to go around in circles, whereas an organism without memory does not fall into that trap. Likewise it is not necessary for natural selection to have any foresight or capacity for analyzing problems for it to be able to advance steadily forwards.

Returning now to the theme of combinatorial games, the problems mentioned are sufficient for introducing the essential idea of metabolic optimization, and looking at the pentose phosphate pathway in the light of it. First I shall treat this pathway as the game defined in Figure 5.2, but afterwards we shall see how the apparently arbitrary rules actually apply to the real biochemical situation.

[1] "Changing things" perhaps gives the wrong impression, implying some conscious intention to test different possibilities. In reality there is no intention, just mistakes, but these have the same effect.

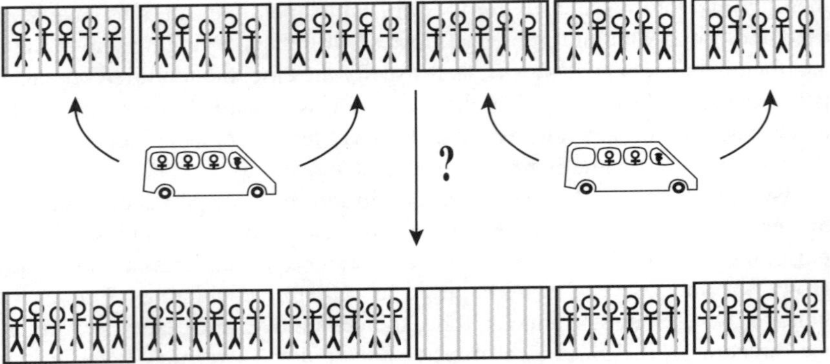

Fig. 5.2 Redistribution of prisoners. The aim of the game is to find the smallest number of steps to redistribute the 30 prisoners into five blocks with six prisoners in each, leaving one empty block. Two or three prisoners must be moved in each step. No block can contain fewer than three prisoners at any moment unless it is completely empty, but a block is allowed to contain more than six prisoners during the redistribution process

We start with a prison camp for dangerous prisoners that consists of six widely separated blocks that house five prisoners each—30 prisoners in total. However, each block is large enough to accommodate more than five prisoners: it can hold six on a permanent basis, or more than that for short periods, during the process of reorganization, for example. The authorities have therefore decided that it is uneconomic to use six blocks for 30 prisoners: one block is to be closed, and each of the remaining five is to accommodate six prisoners.

You, as governor of the prison, have to organize the transfers. As the blocks are widely separated you need to use a car for the transfers, but this presents no problem as you have a suitable vehicle that can accommodate a driver and up to three prisoners. In doing so, you have to take account of the violent behavior of the prisoners, which means that you cannot leave fewer than three together in a block at any time, because two of them left alone will certainly fight, and any one left alone is liable to commit suicide. At this point the problem appears quite easy: you move one prisoner from the first to the second block, then another prisoner from the first to the third block. This leaves three, six, and six prisoners in the first three blocks. Then you do the same thing with the other three blocks, leaving three, six, and six prisoners in these three blocks. You then move the three remaining prisoners from the first block to the other one that contains just three. This leaves the first block empty and each of the other five blocks with six prisoners each, so the problem is solved in just five transfers. Moreover, it is clear that no solution involving fewer than five transfers can exist, because each of the five blocks occupied at the end of the operation must have appeared at least once as the receiving block of a transfer.

Unfortunately, however, you not only have to contend with the genuine constraints in the prison; you also have to follow rules imposed by the bureaucracy responsible for the prison service. The financial director has decided that it is inefficient to use the vehicle to move fewer than two prisoners at a time: you can move two or three at a time, but not one. It is no use arguing that following this rule will make the whole reorganization more expensive. Bureaucrats are impervious to such arguments, and insist that they know best and you just have to obey the rules.

The problem is now appreciably more difficult, and it may not be obvious at the outset that it can be solved at all, even inefficiently. I shall give the solution shortly, but before reading on you may like to reflect on how you might set about solving it. In this way you will arrive at an opinion about how difficult the problem is, and how likely it is that the best solution could be found by trying out different possibilities haphazardly.

We can represent the initial state as 555555 and the desired final state as 666660. You cannot begin by moving three prisoners, as this would produce a block with only two prisoners left in it, so the first step can only be to move two prisoners from one block to another, giving a state that can be represented as 375555 (or something equivalent like 575553). After this there are several possibilities, in which I have written 10 as X to make it a single digit:

Two prisoners moved:	55 → 73	375555 → 373755	
	75 → 93	375555 → 393555	
	75 → 57	375555 → 357555	(synonymous)
	53 → 35	375555 → 573555	(synonymous)
	73 → 55	375555 → 555555	(regressive)
Three prisoners moved:	75 → 48	375555 → 348555	(unphysiological)
	53 → 80	375555 → 078555	
	73 → X0	375555 → 0X5555	

Two of these possibilities are labeled “synonymous,” meaning that the new state is equivalent to the first, and another is labeled “regressive,” meaning that it recreates a state that existed earlier. Clearly none of these three kinds of moves is helpful, and we can ignore them, but it is not obvious which of the other five is best. Another is labeled “unphysiological,” because although not forbidden by the rules as I have stated them, it turns out that, in the biochemical equivalent of the problem, moving three prisoners at a time always involves a block with three prisoners in it; however, in solving the non-biochemical version of the problem we shall allow moves of this kind.

We can continue with each of the five new states in the same way, examining what new states can be created from them, and in the brute-force approach to the problem, the computer programmer’s approach if you like, that is the easiest thing to do. We just write down all the possibilities, following each pathway through until we eventually reach 666660, and select the shortest. To simplify matters you can make a preliminary guess that it is not necessary to pass through a state with more than seven prisoners in the same block, that is, you can ignore any move that leads to a code with an 8, 9, or X in it. If you do this you will find that five seven-step pathways exist, as well as many longer solutions. The number of solutions is actually in finite if we allow solutions that involve aimless cycling around the parts of the network where cycling is possible.

All of the seven-step solutions start with 555555 → 755553, and all finish with 666543→ 666633 → 666660, but they differ as to how they get from 755553 to 666543:

(a) 755553 → 775533 → 745563 → 445566 → 645366
(b) 755553 → 773553 → 746553 → 766353 → 466653
(c) 755553 → 455556 → 473556 → 446556 → 646356
(d) 755553 → 455556 → 473556 → 673356 → 646356
(e) 755553 → 455556 → 635556 → 633756 → 663456

According to the rules defined at the outset these five solutions are equally good, as they have equal numbers of steps. However, is there any sense in which we can say that one of them is "simpler" than the other four? We can answer this by examining the penultimate state 666633, which is the same in all five solutions. This can be regarded as 663 twice over, so that we may ask whether the problem to this point can be regarded as the conversion of 555 to 663 twice over. Solution (e) is plainly of this form, as it consists of 555 → 753 → 654 → 663 carried out first on one 555 group and then on the other. At first sight the other four solutions appear not to fit this pattern, but in reality they do, because they all contain two 555 → 753 → 654 → 663 sequences, but unlike solution (e) they do not complete the first of them before starting the second. It follows therefore that all five solutions are equivalent and we cannot say that any is better than any other.

As I have mentioned, I have preferred to solve this problem by brute force, to better illustrate the sort of approach a computer program or natural selection would use. However, it is worth noting that you can arrive at the right solution much less laboriously by using a little intelligence. First you may deduce that a 33 → 60 step will be needed at some stage to produce the empty block and a block with six prisoners, and you may then recognize that it makes sense for this to be the last step so that the rest of the problem is symmetrical, requiring 555 to be converted twice into 663. All that then remains is to find the simplest way of converting 555 into 663.

Later we shall see which solution has been adopted by living organisms for the equivalent biochemical problem, but before we can do this we need to see how the problem as I have presented it relates to the pentose phosphate pathway. This consists of both oxidative and nonoxidative phases, but we shall consider only the nonoxidative phase, which involves exchanging the carbon atoms of sugars so as to transform six pentose molecules into five hexose molecules, that is, to convert 6C5 into 5C6. The exchanges are brought about by enzymes that transfer a certain number of carbon atoms from one sugar to another. The mechanisms available to the cell are the transfer of two carbon atoms, catalyzed by the enzyme transketolase and the uniting of two sugars (one of them always with three carbon atoms) to make a single one, catalyzed by aldolase.

The problem that poses itself at this point is how to organize these reactions so as to bring about the complete conversion (6C5 → 5C6) in the least number of steps. A little reflection will show that this is exactly the same problem as the one we have already solved, expressed in different words, aside from the additional point that aldolase always uses at least one sugar of three carbon atoms. In terms of the original problem, transketolase is the enzyme that catalyzes moves of two carbon atoms, and aldolase is the enzyme that catalyzes moves of three. The rule in the original game that you could not have a block with one or two prisoners in it derives from the empirical observation that there are no sugars with just one or two carbon atoms. Of course, this is partly a matter of definition: the poison formaldehyde (the main component of the disinfectant formalin) has only one carbon atom, and can be considered a carbohydrate, but it is not considered to be a sugar because it is not used as a sugar by any organism: probably it is too reactive as a chemical to be kept under the degree of control needed for a metabolite. Likewise glycolaldehyde, a carbohydrate with two carbon atoms, exists and participates in some metabolic reactions, but is not regarded as a sugar by biochemists (though it does appear to be regarded as a sugar by astronomers anxious to find extra-terrestrial life).

Let us now express the solution to the combinatorial game in terms of the real biochemical problem: in effect, we found that we could accomplish the conversion of three pentoses into two hexoses and one triose by the following steps:

555 –(TK)→ 753–(TA)→ 654–(TK)→ 636

where –(TK)→ is the transketolase reaction that we previously called a move of two prisoners, and –(TA)→ is the aldolase reaction, previously called a move of three prisoners. As it does not matter which position is which in the sequence 636, we can write it as 663 to bring it into closer relationship with the discussion above, so the three-step conversion be written as follows:

555 –(TK)→ 753 –(TA)→ 654–(TK)→ 663

If each of the two groups of three pentoses is transformed in this way, the two trioses left over can be made into one hexose with a final aldolase step:

33 –(TA)→ 6

Actually, this is not the only seven-step solution that exists, because there is an alternative way of getting from 753 to 663 in two steps that we did not consider because we restricted the discussion to stages with no more than seven carbon atoms in one molecule:

753 –(TK)→ 933 –(TA)→ 663

This involves exactly the same kinds of steps but puts the transketolase step before the aldolase step. Clearly it results in a solution just as short as the first one, but is it any simpler? Apparently not, because it involves all of the same molecules as the first solution (3, triose; 5, pentose; 6, hexose; 7, heptose) with one additional one, a 9-carbon sugar. So it is a more complicated solution, as it not only requires a new molecule not needed in the first solution, but this new molecule must be of a larger and more complicated kind than any of the others. We are left with the conclusion therefore that the original solution is indeed the simplest.

It remains to ask how living cells have solved this problem: which solution is found in real metabolism? If you study the scheme of the pentose phosphate pathway that appeared as Figure 4.2 in Chapter 4, you will see that the answer is that the solution that mathematical analysis proves to be the simplest is indeed the way that pentoses are transformed into hexoses in cells.

Earlier in the chapter we saw that the problem of how to send a tamper-proof message was once thought impossibly difficult, but became almost trivially easy when approached in the right way, and we have a similar case here. Before Enrique Meléndez-Hevia decided to investigate whether metabolism is optimally organized, most biochemists would have doubted whether the question made sense, and would have seen little hope of answering it. Certainly, no one had previously thought of treating it as a mathematical game.

As I discussed in Chapter 1, biochemistry is much the same whether we study it in *Escherichia coli* or an elephant. How did it happen, then, that the right questions were first asked on an island in the Atlantic Ocean and not in one of the main centers of the subject? Probably the crucial step was to realize that the problem could be investigated at all, and a relative isolation from the centers of power allowed calmer reflection, independent of current fashions, than would have been possible in Madrid or Barcelona. Moreover, Meléndez-Hevia's father was a distinguished paleontologist (and his brother is another), and he grew up in a household where problems of evolution were a matter of everyday discussion. As an accomplished amateur composer of music, he also doubtless has less difficulty than most of us in keeping mental track of all the interconnections of the pentoses and hexoses.

Before accepting too readily the conclusion that natural selection has arrived at the best possible solution to an evolutionary problem we need to answer an obvious accusation that the rules as I have given them are arbitrary; rather than examining how nature solved a genuine problem I have apparently just constructed an artificial problem whose solution happens to be the one found in nature.[2] This objection is not altogether unfounded, and there is

[2] That is exactly what happened two centuries ago in the theory of statistics, with results that continue to confuse textbook authors to this day. Karl Friedrich Gauss is commonly credited with proving that the ordinary mean is the best kind of average because it follows from the normal (or "Gaussian") distribution of errors. But in fact he quite explicitly did the opposite, deciding at the outset what conclusion he wanted to reach and then working out what properties the world would need to have for it to be valid.

one rule in particular that cries out for justification. It was all very well in the problem of moving prisoners to invoke the rule that the car could not be used to move one prisoner at a time as a piece of bureaucratic silliness, but such an argument will hardly do for the organization of metabolism, where there is no bureaucracy.

So we need to explain the lack of an enzyme to move one-carbon fragments in a different way. This has to do with the fact that although we have treated all pentoses as a single entity, and similarly with trioses, tetroses, hexoses, and heptoses, the reality is that each class of sugar contains numerous types. In the scheme to illustrate the pathway (Figure 4.2), you can find two different trioses, glyceraldehyde 3-phosphate and dihydroxyacetone phosphate. In the whole pathway, the number of distinct sugars that appear is very much smaller than the total numbers of sugars in the various classes. If all possible trioses, tetroses, pentoses, hexoses, and heptoses appeared in the pathway it would be vastly more complicated than what is shown, and that would not just be a problem for students; it would also be a problem for the cell, which needs to keep track of all the different components. Keeping the total number of participants within manageable limits requires some degree of specificity in the enzymes that catalyze the reactions, accepting some sugars as possible substrates, but not others. When this is taken into account, together with the fact that chemical principles do not allow an enzyme to have a completely arbitrary specificity, it turns out that an enzyme capable of transferring one-carbon fragments could not be specific enough to generate a complete pathway as simple as the one that actually exists.

I have taken some trouble to present the solution to the problem in a way that is as simple as possible to follow, first showing that brute force will inevitably lead to the right answer eventually, albeit with a lot of work, and then showing that with a little logical analysis you can arrive at the same result much more quickly. All of this may give the impression that the fact that cells have adopted the simplest solution has no importance, because the best solution may appear obvious once it has been pointed out. In reality it is not obvious at all, as you may readily check by setting the combinatorial game as a problem for someone who has not encountered it before and does not know what the solution is. In practice such a person will normally flounder around for a considerable time before finding the best solution. Remember too that by presenting the public-key cipher problem in terms similar to those used in the problem of the farmer and the river, just after discussing that problem, we made its solution seem simple and obvious, although in historical reality the solution escaped some of the greatest experts in cryptography. Almost any great and original discovery can be labeled "obvious" once its solution has been found, explained, and understood.

Once we accept that living organisms have found the best solution to a nontrivial problem we need to ask what steps were followed during evolution to arrive at it, as we cannot easily suppose that the primitive organism that first adopted the pentose phosphate pathway just happened to organize it in the best possible way. Human players use their intelligence to improve their strategy, because they reflect and learn lessons for the next time, but this approach is not open to a mechanism that has no intelligence and no foresight. Answering that natural selection works by tinkering, rejecting the modifications that fail, and selecting the ones that are better, is inadequate, because even if it is true it leaves some important questions unanswered: how does an organism decide that one way of organizing the pentose phosphate pathway is better than another, and in what sense is a simpler solution better than a complicated one? I shall try to answer these later in this chapter, but first we need to consider whether the success in the pentose phosphate pathway is just a one-off chance result, or whether there is any evidence that metabolism in general is optimized, rather than just one pathway that happens to have been studied.

Photosynthesis, the process that green plants use for harnessing the sun's energy, provides another example. It involves a metabolic pathway called the *Calvin cycle*, which, like the pentose phosphate pathway, contains a portion that provides a mechanism for converting one set of sugars into a different set. However, the requirements are sufficiently different that it is not just the same problem in a different guise: it is a different problem, and the fact that the problem of the pentose phosphate pathway has been solved by living organisms gives us no reason to expect the corresponding problem for the Calvin cycle to have been equally solved unless we accept the existence of a general tendency in evolution to find the simplest solutions to metabolic problems. If this problem is expressed in the same terms as those we used for the pentose phosphate pathway, the objective is to convert 12 trioses (three-carbon sugars) into one hexose (six carbons), and six pentoses (five carbons each).

Applying the same rules as before (transfers restricted to fragments containing two or three carbon atoms; no sugars with fewer than three carbon atoms), the pathway as it exists in plants has the organization illustrated in Figure 5.3. You may like to search for a simpler way of achieving the result, and, if you do, you will find that you cannot, so the Calvin cycle, like the pentose phosphate pathway, is optimized in living cells. This gives us good reason to believe that such optimization may be a general characteristic of metabolism.

After noticing that evolution has arrived at the simplest ways of organizing these pathways, we may feel that it is obvious that the simplest solution is the best solution and that no discussion of the advantages of simplicity is necessary. This is naive, however, because if we look at metabolism as a whole we can find many examples where it is not obvious at first sight that the simplest approach has been adopted.

One of the best known pathways is glycolysis, the anaerobic (oxygen-independent) part of the mobilization of glucose: this is often the first pathway that students are taught, and often the one that they know in most detail; it usually occupies the central part of a metabolic pathways chart, the part that is printed most prominently. However, examination of the reactions of glycolysis reveals a peculiarity. Glycolysis begins with the conversion of glucose to

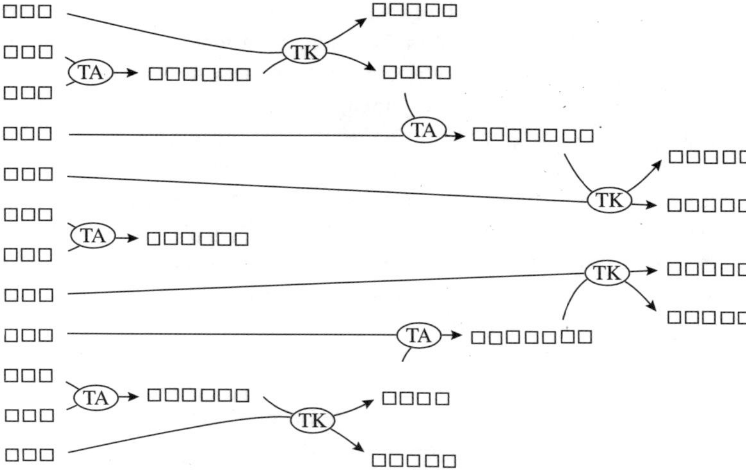

Fig. 5.3 Solution to the Calvin cycle game. The aim of the game, similar to the one shown in Figure 5.2, is to redistribute squares two or three at a time such that 12 groups of three squares are converted into one group of six squares and six groups of five

glucose 6-phosphate, that is to say by the transfer to glucose of a phospho group from ATP, the energy currency of the cell. A little later in the pathway another phosphorylated sugar, fructose 6-phosphate, receives a second phospho group in the same way, making it into fructose 1,6-bisphosphate. Yet in the later steps of glycolysis these phospho groups are removed, regenerating ATP, and additional ATP is made with inorganic phosphate ion as the source of the phospho group.

What is the point of transferring two phospho groups from ATP in the early stages of glycolysis, only to transfer them back again later on? Surely it would be simpler and more efficient to dispense with these apparently unnecessary steps? This is a little like asking the point of requiring passengers in a moving vehicle to attach their seatbelts even though we know they will have to detach them again at the end of the journey! Actually the phosphorylation allows the process to be forced to go in the required direction, and makes it easier for the regulatory mechanisms to determine the rate at which it proceeds. In addition, the phosphorylated sugars cannot cross biological membranes without specific mechanisms to take them across, and so they cannot leak into parts of the cell where they are not wanted.

It follows, then, that in considering whether a particular pathway is optimized for simplicity we must avoid posing the question in excessively simple-minded chemical terms: apparently pointless steps may fulfill necessary functions. Incidentally, all of the sugars we considered earlier in the chapter are also phosphorylated, but, as they do not become unphosphorylated in any of the reactions considered, we could discuss the organization without taking account of the phosphorylation.

We cannot assume that simplicity will be selected because of the intellectually pleasing results that it produces. To understand why simplicity should be selected we must understand why it is advantageous to an organism to have a simpler pathway, why such an organism will live longer and have more descendants than a rival with a more complicated arrangement. In short, the answer is that a shorter pathway is more efficient for converting metabolites into other metabolites than a longer pathway: it uses less materials, whether these are chemicals stored as intermediate metabolites in the pathway, or protein needed to make the enzymes that catalyze the reactions; more important, it is faster. Other things being equal (and this may be an important qualification, as there is no very strong reason to be sure that other things will actually be equal), a pathway catalyzed by two enzymes will transform a given concentration of starting material into product faster than a pathway of three enzymes with the same kinetic properties as those of the first. Let us examine this in terms of a simple experiment illustrated in Figure 5.4, which is easy to set up with equipment available at home or from a hardware shop.

Take two identical tanks, each with an outlet near the bottom to which a section of plastic tubing can be attached in a watertight way. For one tank, connect a tube of a meter or so in length, draining into a sink; for the other tank do the same thing, but use a tube about twice as long. Place clips on both tubes near their bottom ends so that the flow can be switched on or off at the level of the sink. Then fill both tanks with water, place them at exactly the same height above the sink, allow the water to flow until each tube is full of water and then cut off the flow. Ensure that both tanks are filled to the same level, and that the outlets of both tubes into the sink are at the same level. If all this is done correctly the water pressure at the outlet of each tube should be the same, and the amount of water in each tank should be the same, so the only difference between the two cases is the length of tube that needs to be

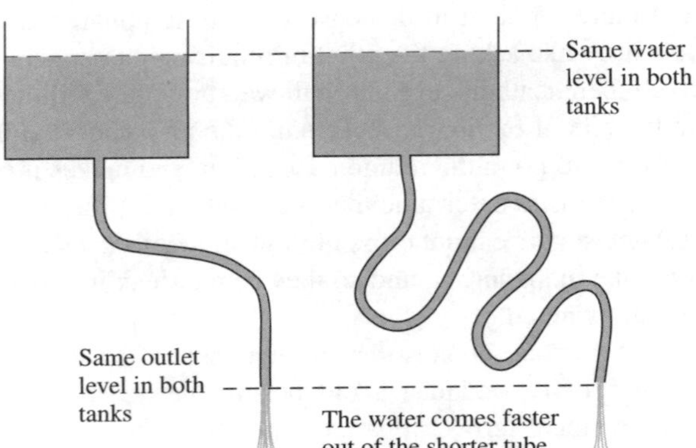

Fig. 5.4 Emptying two tanks. Two identical tanks are filled to the same level with water, and are connected to exit tubes of identical internal diameter ending at the same level but of different lengths. The tank with the shorter tube empties more quickly

traversed by the water on its way from the tanks to the sink. Now remove the two clips simultaneously. Which tank will empty first?

Two reasonably plausible answers may be given: either there will be no difference, or the tank with the shorter tube will empty first. (Not many people suggest the third possible answer, that the tank with the longer tube will empty faster, and this is certainly wrong.) Whichever you think, you are quite likely to think that it is obvious, and that the other possibility is absurd, but if you cannot make up your mind then do the experiment. In fact it is easy to do, and if you do it you will find that the second answer is correct: the tank with the shorter tube will empty more quickly.

If you do not believe this, please do the experiment. Do not do what a colleague of mine did after reading about it. He told me that the experiment did not work as described, so I was worried, because at that stage I had not actually tested it but just believed Meléndez-Hevia's account of it in the book that I mentioned in Chapter 4. I thought my colleague meant that he had tried it for himself and found it to fail. But no, that was not what he had done. Despite being an experimental biochemist (whereas I am sometimes accused of being an armchair biochemist, too much interested in theory and not enough in the experimental realities), he consulted a physicist who knew about hydrodynamics. The physicist did not do the experiment either, but made some calculations on the back of an envelope and reported back that the flow would be exactly the same in both tubes! Since then I have done the experiment (it only takes a few minutes) and found that the tank with the shorter tube emptied more quickly, as I said above. So much for theory when inappropriately applied.

The tank experiment also illustrates—in a way that is so obvious that simply describing the arrangement is sufficient—the two other advantages of the shorter tube in addition to flow rate: it uses less material for achieving the connection between tank and sink, less plastic for the analogy, and less protein for the metabolic pathway; less material is tied up in the pathway itself during the process, less water in the tube at any moment, and less chemical resources in the pathway intermediates. This type of consideration played a part in the decline of the canals and the replacement of their functions by railways in the second half of the nineteenth century. Canals were very efficient for transporting nonperishable goods, and much cheaper in principle than railways. Moreover, for goods like coal that are not subject to sudden wild fluctuations in supply and demand, they could ensure that the output rate always matched the input rate, so their inherent slowness might seem to have little importance. However, the slowness meant that at any particular moment a large amount of material was in transit, so the supplier would suffer if it had not been paid for, and the buyer would suffer if it had been paid for but had not yet been delivered.

Returning to the pentose phosphate pathway, it follows that an organism with a shorter sequence of reactions to accomplish the conversion will have some economic advantages over a rival that uses a longer sequence to achieve the same result. This much is clear, but what is perhaps less clear is the evolutionary route by which a less-efficient system can be transformed into something shorter and better. Once a given set of reactions is in place, it may seem to require major reorganization to replace the chosen reactions by a different set: new catalytic activities need to evolve, the now redundant older enzymes need to be eliminated, and so on. All this requires some improbable events to occur, whereas somewhat less-improbable events may improve the performance of the organism that retains the longer pathway. For example, mutations that alter the structures of proteins occur from time to time in all lineages, and although many of these are harmful or at best neutral, occasionally these can result in an enzyme that catalyzes its reaction more effectively than the unmutated enzyme.

As these events are purely random, and have no special tendency to improve performance in organisms that are already improving their metabolism, it is perfectly possible that an organism that is on its way toward improving its metabolic design may compete with another that retains the less-efficient design but happens to have mutated some of its enzymes to be more effective catalysts. In such a competition, the organism with a less-efficient design may still be able to live a more economic life and may overrun the other. However, just as there is no reason for the improving organism to be specially favored with valuable new mutations, there is equally no reason for the opposite to be true. On average, therefore, and given enough time, none of these chance effects will favor one organism over another, and ultimately we may expect the more-efficient design to win. Thus it is that today we find that, in all organisms that have been studied, the pentose phosphate pathway and the Calvin cycle are arranged to use the smallest number of reactions possible for the conversion.

Actually the organism that retains the less-efficient pathway has available to it an easier way of increasing fluxes than selecting rare mutants that improve the catalytic activities of the component enzymes: instead of improving the enzymes it can simply make more of them, because, in general, more catalyst means a higher rate. However, as a way of competing with an organism that has found a more-efficient pathway this is ultimately ruinous: more enzyme means more protein, and protein is expensive; an organism cannot just make unlimited amounts of protein without considering the cost in terms of dietary sources of aminoacids and the energy resources needed for assembling them into proteins. Some enzymes, indeed, are already present in such large concentrations that it is hard to imagine that their amounts could be increased. Ribulose bisphosphate carboxylase, an enzyme necessary for photosynthesis, is the most abundant protein on earth, and accounts for about half of all the protein in the leaves of green plants. Even a 5% increase in its concentration would imply severe strains on the supplies of many other proteins needed for the health of the plant. This parallels the example in Chapter 3, where we saw that the blood of a mammal is packed so full with hemoglobin that there is no room for any more. More generally, just increasing the quantities of enzymes in an inefficient pathway is not an option for competing with an organism that has found a better one.

The pentose phosphate and Calvin cycles are, of course, just two rather similar metabolic pathways out of many known to biochemistry. For the moment, therefore, it would be dangerous to generalize and claim that all of metabolism can be analyzed in the same sort of way and is correspondingly optimal. Clearly more work will have to be done, but the idea that metabolism has evolved to an optimal state needs to be regarded as a serious possibility. What of extraterrestrial life, if it exists? I shall discuss this in a general way in Chapter 11, but meanwhile there is no reason to doubt that if optimality has been achieved in terrestrial metabolism it should also be a feature of metabolism elsewhere in the universe, though the detailed chemistry may well be different.

6
The Perfect Molecule

The pursuit of perfection, then, is the pursuit of sweetness and light . . . He who works for sweetness and light united, works to make reason and the will of God prevail.

Matthew Arnold, *Culture and Anarchy*

Perfection is the child of Time.

Bishop Joseph Hall, *Works* (1625)

When I chose a title for this book I did not realize that I was quoting Matthew Arnold, still less that placing it in context would make it rather inappropriate for the idea that blind chance operating over several billion years has produced the appearance of perfection in many aspects of biochemistry that we can recognize today. The idea is conveyed rather better by the second quotation above, though given that it comes from the hand of a bishop he probably had something different in mind from what concerns us here, but no matter. In the preceding two chapters I have discussed Enrique Meléndez-Hevia's demonstration that the metabolic pathways he examined are organized in living organisms as well as they possibly could be. He has also been interested in other questions of optimality in biochemical systems, and here I discuss one of these, the structure of the storage carbohydrate glycogen.

In principle, we could ask of almost any molecule selected for a prominent biochemical role whether its structure is really well suited to its function, or could be designed better. For most structures, however, it is not obvious how to put the question in a concrete testable form. Until we know a great deal more than we do now about how the details of protein structure are related to the fine details of enzyme activity, it will be difficult to know whether the enzymes we have are the best we could possibly have. Certainly, there have been reports that artificially modified enzymes can have higher catalytic activity than their natural counterparts (though not usually by large factors) but this does not prove that they are "better," because a real enzyme has more to do than just catalyze a reaction. It has to have the required degree of stability: not so stable that it fails to disappear when it is unwanted; not so unstable that it needs to be constantly synthesized in order to replace the molecules that are constantly being lost. It must be capable of being recognized by other macromolecules that interact with it, either to modify its catalytic activity (or their own) or to move it according to metabolic signals from one location in the cell to another. At the moment, proteins are just too complicated in structure and function for us to have any confidence in estimates of whether they are optimally designed or not.

There is, however, a macromolecule with a function that can be expressed very precisely, that is built from a vast number of copies of a single kind of building block, and that needs to satisfy functional constraints that can be defined precisely. It is possible, therefore, to assess the extent to which the structure that exists in nature is optimal. This is the storage carbohydrate glycogen, and Meléndez-Hevia has studied its structure very carefully from the point of view of optimization.

Glycogen is sometimes called *animal starch*, and that is a good name because it is similar to starch in structure, and its function in animals is similar to that of starch in plants. It allows considerable quantities of glucose to be stored in liver and muscle in such a way that it makes no detectable contribution to the osmotic pressure of the blood or the cell water, but still allows rapid access to the glucose when it is needed. The need for rapid access is clear enough, and I shall come back to it shortly, but what is osmotic pressure, and why is it important that glycogen does not contribute to it?

Before trying to answer these questions I shall look at a more familiar kind of pressure, such as the pressure of gas that keeps a car tire inflated or prevents a balloon from collapsing. This pressure depends on the *number* of molecules in the gas (and not, for example, on the weight of the gas), and forcing more molecules into a confined space causes the pressure to increase. This can be done mechanically, as for example when a car tire is inflated, causing the rubber to be stretched as a response to the energy input. It can also be done by boiling a confined liquid, as when a sealed container explodes after being left on a gas burner, and it can also be done chemically, as in the explosion of a substance such as nitrocellulose, when a small number of nitrocellulose molecules in the solid state are converted almost instantaneously into a very large number of gas molecules—mostly carbon dioxide, water vapor, and nitrogen.

Osmotic pressure is similar in origin and can have similar catastrophic effects if not properly regulated. Although it is less obvious in everyday life than other sorts of pressure, it does have some easily observable effects. For example, it causes flowers to take up water from a vase. A plant fills with water from its roots because of *capillary action*, an effect due to *surface tension*, which has the even more familiar consequence that water can wet objects immersed in it: when water touches your hand, for example, it does not fall right off leaving your hand dry, but tends to stick to it. However, although capillary action can fill a plant with water—even a tall tree—it does not generate a continuous upward flow: if capillary action were the only process, it would stop once the plant was saturated with water. What prevents it from stopping is osmotic pressure. When water evaporates from the leaves, the amount of water left in the leaf cells decreases, and all of the material dissolved in it becomes more concentrated. Now, just as the pressure of the air in your car tyre depends on the number of gas molecules inside it, the osmotic pressure in a water solution depends on the number of molecules dissolved. Even the numerical details are almost the same: a given number of molecules dissolved in 1 liter of liquid make almost the same contribution to the osmotic pressure as the same number of gas molecules would to the gas pressure if pumped into a closed container of 1 liter. So when water evaporates from a leaf, the osmotic pressure in the leaf cells increases, and the only way the plant can balance the pressure is by bringing more water from elsewhere.

A less-familiar example, but one that is very important for treatment of patients in hospitals, is provided by the red cells in the blood, which burst when blood is mixed with water. This bursting is directly due to osmotic pressure, and people have occasionally died after being injected or perfused with water. This effect is obvious if you examine the blood and the blood–water mixture under a microscope, but not with the naked eye. There is, however, a similar phenomenon that is very easy to see in your kitchen. If you boil potatoes in *very* salty water (add about quarter of a kilogram of salt to water that is just enough to cover the potatoes in the saucepan), then they will not become waterlogged even if you boil them for half an hour or more, but instead will acquire the character of baked potatoes cooked in the oven. In other words they will *lose* water during cooking, not absorb it, because the osmotic pressure of the water is much higher than it is in the cells of the potatoes. Considerations of this kind explain why the fish of the sea need to protect themselves with waterproof skin: despite the apparent abundance of water all around them, it is so salty that they would rapidly become dehydrated if water could flow freely through their scales. Incidentally, potatoes cooked in the way I have described are called *papas arrugadas*, and are popular, appropriately enough, in Tenerife, the island where Meléndez-Hevia works.

Osmotic pressure is crucially important in systems containing semipermeable membranes, and although these do not feature very much in everyday life they are extremely widespread and important in the bodies of living organisms. A semipermeable membrane is a divider between two liquids that small molecules like water can pass through, whereas large ones like proteins cannot. What happens with medium-sized molecules depends on the specific characteristics of the particular membrane considered, but for the purposes of this discussion I shall assume we are dealing with a membrane that is permeable only to solvent molecules like water and not to any of the molecules that are dissolved in it. Connecting two children's balloons inflated to different extents causes air to flow from one to the other until the pressures are balanced, and putting two solutions in contact via a semipermeable membrane causes water to flow across the membrane from the solution with lower osmotic pressure into the solution with higher osmotic pressure.

In principle, this flow continues until the osmotic pressures are balanced, but if one or both of the compartments in contact is sealed, and if the sealed compartment has the higher osmotic pressure (as with a red blood cell suddenly diluted into pure water) it will burst when its membrane can no longer withstand the increase in volume. A healthy red blood cell has rather a flat structure, and its volume can increase quite substantially before it bursts; nonetheless, it inevitably bursts in pure water as the osmotic pressure of pure water is zero, whereas its own osmotic pressure can never fall to zero no matter how much water flows in. In principle, the same sort of thing could happen to any cell immersed in pure water, but many (most plant cells for instance) are much more protected than red cells, because they are encased by strong protective walls that make up for the structural weakness of the membranes.

The point of this digression about osmotic pressure is to emphasize that, although most people can pass their entire lives without ever becoming conscious of its existence or noticing its effects, it does have a crucial importance inside the living body and it is essential to guard against its potentially catastrophic effects. Membranes are everywhere in the body, and the osmotic pressure needs to be balanced across every single one of them. Consider what happens to someone who rapidly eat three 5-g glucose tablets (prescribed to diabetic patients, but also often recommended to healthy people as a source of "energy" by unscrupulous advertisers). Fifteen grams of glucose is about one-twelfth of a mole, a mole being, roughly speaking, the amount of a substance that contains the same number of molecules as there are atoms in a gram of hydrogen.

Measuring quantities in moles may seem a little bizarre if you are more used to weights or volumes, but it makes a lot of sense in chemistry and biochemistry, where we are often more interested in how many molecules there are than in how much they weigh, or how much space they occupy (though these can be important as well). As I have mentioned, osmotic pressure is determined primarily by the number of dissolved molecules, so it is especially important in this context to measure quantities in moles.

Returning to the three glucose tablets, if the 15 g became instantaneously dispersed over the whole volume (about 4 liters) of the blood, it would cause an increase of about one-fiftieth of a mole per liter. This would be a tremendous increase, as the normal concentration of glucose in the blood is about a quarter of this, around one two-hundredth of a mole per liter. It would bring about a correspondingly large increase in the osmotic pressure of the blood—not a fourfold increase, certainly, as there are many other things dissolved in the blood as well as glucose, but about a 15% increase nonetheless—enough to cause serious problems for the person.

Fortunately the effects of eating too much glucose in a short time (or adding too much sugar to your tea or coffee) are not as dramatic as this calculation may suggest, in part because glucose taken by mouth does not pass instantaneously from the stomach to the bloodstream. Nonetheless, it does pass quite fast, and to avoid excessively perturbing the sugar concentration and osmotic pressure of the blood it is essential to have an efficient and rapid way of converting glucose into something that does not contribute to the osmotic pressure. This "something" is glycogen, which can contain more than 50 000 glucose units linked together in a particle known as a *beta-particle*. In muscle the beta-particles are stored as such, but in liver, with which we are mainly concerned here, they are organized into larger assemblies called alpha-particles, each of which consists of several beta-particles.

As the osmotic pressure depends on the number of particles, and not on their size, this system allows the liver to store the equivalent of a huge concentration of glucose, up to 400 mmol/liter, or 80 times the concentration in the blood. (This would be enough to produce an osmotic pressure of around 10 times the pressure of the atmosphere, the sort of water pressure a diver feels at a depth of about 100 m.) However, storing a large amount of glucose is not by itself enough: the structure needs to be capable of being made and broken down fast, so that the glucose is available when needed. This is especially important for the glycogen stored in muscles, which may need to convert much of it into mechanical work in a very short time. In the liver, speed of release is much less important, but it does no harm to use essentially the same structure in the liver as is needed in muscles.

To allow rapid mobilization, the whole surface of the glycogen molecule needs to have many points for attack by the enzymes that remove glucose from it. We can think of these enzymes like animals that graze by eating the leaves from a tree and can only reach the exposed ends. (I am thinking here, of course, from the point of view of the grazers, not from that of the tree, which has certainly not adopted a particular form in order to be eaten most efficiently.) To expose the maximum number of leaves, the tree needs to have a highly branched structure, and that is also how glycogen is arranged, with more than 2000 *nonreducing ends* (as they are called), available on the surface of a beta-particle.

From this description, it is clear that glycogen has a good structure for fulfilling its function, but that is not the same as saying that it is optimal. How can we be certain that tinkering with the structure—making the branchpoints closer together or further apart than they are in the real structure, for example—might not result in something even better? To establish this we need to make some quantitative comparisons, and we need to base them on the real structure of glycogen, not the fantasy structures illustrated in most textbooks of biochemistry.

From the time when glycogen was first discovered by Claude Bernard in the nineteenth century, it was clear that it was a device for storing glucose, and that it must therefore consist mainly of glucose molecules linked together. However, the structure of glucose is such that there are numerous different ways of linking the molecules together to make a polymer. Cellulose, for example, the principal structural material for making plants, is also a polymer of glucose, but it has a different structure from glycogen and starch and has different properties: for example, starch is very easily digestible in the mammalian gut, whereas cellulose is completely indigestible. There are in fact two different kinds of links between glucose units in glycogen, which are called alpha (1 → 4) and alpha (1 → 6) links. The numbers here refer to the six different carbon atoms that each glucose molecule contains, so alpha (1 → 4) means that the atom labeled 1 in one molecule is attached to the atom labeled 4 in another glucose molecule. Another kind of link, called a beta link, occurs in other biologically important molecules such as cellulose, the main structural material in plants. Although the two kinds of links are superficially rather similar, they lead to dramatically different properties: starch and glycogen on the one hand, and cellulose on the other, are all made primarily of (1 → 4)-linked glucose units, but you can nourish yourself by eating potatoes (made mainly of starch) but not by eating paper (made mainly of cellulose).

As glycogen has both (1 → 4) and (1 → 6) links we see immediately that carbon atom number 1 is used in every link. Any glucose unit can therefore be used only once at the left-hand end of an arrow, but it can appear once or twice at the right-hand end. This allows a branching structure to be built, but not a more complicated sort of network with lines joining as well as separating. Within the glycogen structure there are in fact large numbers of linear arrays of several glucose units joined by alpha (1 → 4) links, which come in two varieties, known as *A-chains* and *B-chains*: the B-chains are branched and the A-chains are not, and the branches in the B-chains are produced by alpha (1 → 6) links, each such link making the start of an A-chain. In addition to these two main kinds of chains, there is a single example of a C-chain, which is similar in structure to a B-chain except that it is attached to a small "primer" protein called *glycogenin* (small compared with the glycogen molecule but, like any protein molecule, large compared with a glucose molecule). This glycogenin is located at the center of the entire structure.

In Figure 6.1 glycogenin appears as a circle labeled G and the three kinds of chains are labeled A, B, or C as appropriate. Notice that all the chains are of the same length (have the same numbers of glucose units) and each B-chain is normally the starting point for two other chains. Notice also that this branching system produces a series of layers, as illustrated by the gray circles in the background. In the complete beta-particle there can be as many as 12 layers, but this would be almost impossible to draw (50 000 glucose units!) in an intelligible way, so only the five innermost layers are shown.

Understanding of this structure did not come instantaneously, of course, and unfortunately a wrong structure became thoroughly entrenched in the textbooks before the true one was known. In the early years there were numerous speculative structures that were consistent with the sparse knowledge of the facts. However, only one of these survived the discovery in the 1950s that there were approximately equal numbers of A and B-chains, and this was one that had been proposed by Kurt Meyer and Peter Bernfeld in 1940. It is not illustrated here, because it is wrong, though it is the only one you will find illustrated in almost any standard biochemistry textbook.

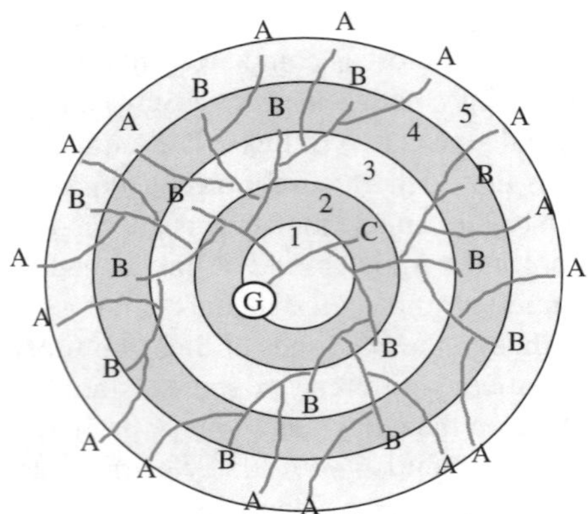

Fig. 6.1 Structure of glycogen. The diagram illustrates the structure of the storage carbohydrate glycogen established by William Whelan and colleagues. It is *not* the same as the (wrong) structure illustrated in most biochemistry textbooks. There are three kinds of chains of glucose units labeled A, B, and C, together with a small protein called glycogenin at the center (labeled G). A-chains are linear, but B and C-chains are branched. There is only one C-chain, which differs from the B-chains by being connected to the central glycogenin

While writing this book, I checked in 10 current and widely used university textbooks of biochemistry for science students. Not a single one of them showed the correct structure. Seven of them plainly showed the obsolete Meyer–Bernfeld structure; of these, five mentioned glycogenin, albeit in another context, but the other two ignored it. Of the others, two did not show sufficient information for me to know whether their text referred to the real structure or not, and the last drew a complete structure in such a vague way that I could not be certain what structure it was supposed to be, but most likely the Meyer–Bernfeld structure was intended there as well. All is not necessarily lost, however, as I do know of one textbook intended for medical students that shows the structure of glycogen correctly.

This correct structure (initially without glycogenin, which was discovered later) was established by William Whelan and his group in 1970—long enough ago for the information to have filtered through to the authors of textbooks. There is, incidentally, no controversy about this: the structure determined by Whelan and colleagues is accepted by everyone who works on glycogen. The two essential points for analyzing whether the structure is optimal or not are that all of the chains are of the same length and all of the B-chains, except those at the outermost layer, have two branchpoints, neither more nor fewer. Instead of the rather regular arrangement shown in the illustration, the obsolete structure has a more haphazard appearance, with considerable variation in chain lengths and with more or less than two branchpoints in each B-chain. The wrong structure is, incidentally, more difficult to draw than the correct one, and more difficult for students to remember, so it is doubly hard to understand why textbooks persist in showing it in preference to the real structure.

We are now in a position to analyze whether glycogen could be improved for its function by modifying the structure. There are three characteristics that we would like to maximize: the number of chains on the surface available for attack by the degrading enzymes, the number of actual sites for attack, and the total number of glucose molecules stored in the complete structure when it reaches its maximum size. In addition, there is a fourth characteristic that we would like to minimize, the total volume of the model. This last is, like the need for rapid mobilization, especially important in muscle. It may not matter very much if the liver is somewhat larger than it would be in a perfect world, but every unnecessary milliliter devoted to muscle represents a significant cost in terms of less-efficient energy generation. Taking account of all these characteristics we can define a mathematical function that we would like to optimize, so that we can say that the best structure is achieved when the function is maximized.

Of course, glycogen is not a protein, and so there is no gene as such for it. Instead there are genes for the enzymes that make glycogen, and the normal evolutionary mechanisms can act on these enzymes so as to vary the final structure of the glycogen that they make. They can vary, for example, the numbers of glucose units in each chain and the frequency with which new branchpoints appear.

So far as the degree of branching is concerned, the conclusions are very clear. With fewer than two branchpoints in each B-chain the molecule would not fill the space available as it became larger; with more than two it would fill it up so quickly that it would run out of space for adding further layers at quite a small total size. A structure with two branchpoints per B-chain is the only one that allows efficient use of the space available. If this is fixed at two, there is then just one variable that can be adjusted by natural selection, and that is the length of each chain. This cannot be less than four glucose units per chain, as any shorter chain would not have room for two branchpoints, and at exactly four the structure is very far from optimal with very poor capacity for storing glucose and making glucose available at the surface. As the chain is made longer the quality function initially increases very steeply, reaching a rather flat maximum at a chain length of 13, and then declining slowly as the chain length becomes still longer.

It follows that if the glycogen structure has truly been optimized by natural selection, we should find it with two branchpoints per B-chain and a chain length in a range of about 11–15. This is exactly what we do find in many organisms—mammals, fish, birds, invertebrates, bacteria, protozoans, and fungi. The only important exception is an interesting one: oysters have a substantially suboptimal glycogen structure with a chain length that varies wildly from 2 to about 30, averaging about 7. The oyster is an animal that has not changed morphologically since the Triassic era, that is to say in 200 million years. It is generally recognized to be a "living fossil," and although we cannot assume that a conservative morphology necessarily implies a conservative biochemistry, it does appear that the oyster has not kept up with progress in optimizing its glycogen. Given that it does not lead a very exciting life, remaining fixed to a rock and rarely having occasion to make rapid or sudden movements, it is quite plausible to suppose that the selective pressure to improve its glycogen has been feeble compared with that felt by other species. Still, fungi do not lead very exciting lives either, so we should not get carried away by this argument, and it may be that oysters have some as yet undiscovered reasons for possessing a deviant glycogen structure.

Enrique Meléndez-Hevia does not accept the view of some paleontologists that it was mainly luck that decided which species escaped the major extinctions, like the one that caused the dinosaurs to disappear and to be replaced by mammals as the principal large animals on the earth. He believes that when the full facts are known we shall find that the lineages that became extinct did so because of a major inefficiency in their biochemistry that made them unable to compete when it became necessary. He thinks that by studying the few species that escaped the extinction—crocodiles, for example, in the case of the dinosaurs—we may discover what defect the extinguished group had, for example, that they had a suboptimal pentose phosphate pathway. In the case of glycogen, he argues that the oysters may be just such a remnant of an earlier group of animals that did not succeed in evolving a glycogen molecule that was good enough for modern life.

One feature of the glycogen structure (in animals like mammals, not in oysters) that I mentioned but did not develop is that the maximal structure has 12 layers. Why does it stop there? Is the number of layers optimal? Is there a mechanism to ensure that the molecule grows to that size and no more? In his celebrated and in many respects excellent textbook *Biochemistry*, Lubert Stryer says that elongation stops when the enzymes responsible for elongation are no longer in contact with the glycogenin, and he describes this as "a simple and elegant molecular device for setting the size of a biological structure." However, he does not explain how a small protein molecule buried deep inside a molecule more than 25 times as large can remain in contact with enzymes working on the surface, and without this we can only regard this explanation as a fantasy without any basis in reality.

In fact the maximum size is determined by the branching number and the chain length. With a branching number of two the space is filled more and more tightly as the molecule expands, so that the inner layers are rather loosely packed and the outer ones very tightly packed. This inevitably means that the molecule cannot grow forever, and modeling in the computer shows that although a molecule with 13 layers could exist, it would have a very smooth surface with around two-thirds of it occupied by glucose. This might be good if the aim were simply to build the biggest possible molecule, but, in the living organism, enzyme molecules, which are much larger than glucose units, need to have space to work on the surface, either adding more units while it is growing, or taking them off when it is being degraded. A hypothetical 13-layer glycogen would not leave enough space for enzyme molecules to work and so the molecule cannot grow that large.

7
Fear of Phantoms

Jim was a little more brash, stating that no good *model ever accounted for* all *the facts, since some data was bound to be misleading if not plain wrong. A theory that* did *fit all the facts would have been "carpentered" to do this and would thus be open to suspicion.*

Francis Crick, *What mad pursuit* (1988)

Chapter 3 ended with a promise to devote a later chapter to an example of a Panglossian absurdity that illustrates how biochemists have sometimes believed things that were too good to be true, finding adaptations where no adaptation exists. The time has come to make good this promise, and I shall begin by examining why biologists cannot apply the same criteria as physicists for deciding how much trust to place in their theories.

Quantum electrodynamics is the part of physics that deals with the electrical interactions between fundamental particles such as electrons. It leads to the conclusion that the magnetic moment of an electron should differ from the value given by a simpler theory by a factor known as *Dirac's number*. Richard Feynman told us in his book *QED* that the experimental value for Dirac's number was 1.001 159 652 21, whereas theory gave a value of 1.001 159 652 46. So the measured value of the magnetic moment agreed with the theory to better than one part in a billion. In the years since Feynman's book was written, the agreement has probably improved. But, no matter: this is good enough for now.

When physicists obtain such an impressive agreement between experiment and theory they are naturally delighted, and take it to mean that their theory is on the right lines. Biologists, by contrast, should be deeply suspicious: all our experience tells us that few, if any, biological numbers can be measured as precisely as this, and even if they can be measured precisely, values for different organisms never agree exactly because sampling variations and all sorts of complications lead to differences between reality and expectation that are never less than a few percent. For example, if theory teaches us that the sizes of the testicles of the males of different animals ought to be inversely correlated with the degree of monogamy of the species, we might test this by measuring the testicles of numerous animals and plotting them against some measure of

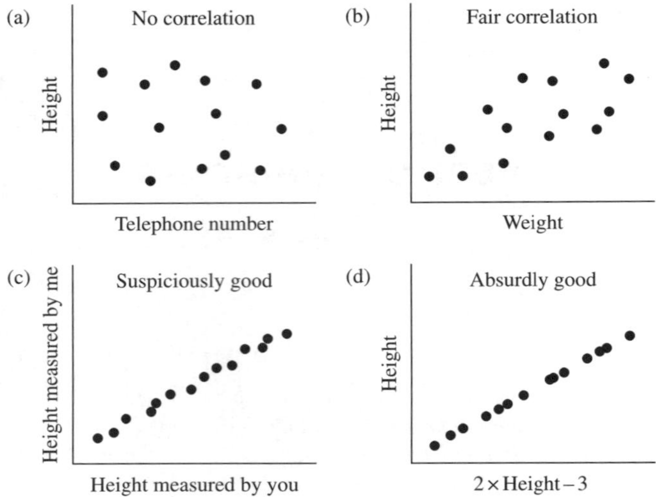

Fig. 7.1 Correlation between measurements. When two different properties are measured for a set of entities, one can plot one of them against the other. (a) If there is no relation between them at all, the points should be scattered with no evident pattern. (b) If the correlation is typical of quantities measured in biology a definite trend should be evident. (c) An excellent correlation is, in biology, typical of the agreement between two sets of measurements of the same thing. (d) A perfect correlation is most likely to arise from one set of measurements transformed mathematically to give the impression that two different things have been measured

monogamy. If we find that the points on the resulting graph are scattered without any hint of any relationship, as in panel (a) of Figure 7.1, we should reject the hypothesis, whereas if they are somewhat scattered but still lie within 20–30% of a trend line, as in panel (b), we should think the degree of agreement quite good.

However, if the agreement between numerous species is within 1–2%, as in panel (c), we should start to worry: this is looking too good to be true; surely we cannot have measured the values as accurately as this? Even if we have, can this really be the whole story? Surely there must be something more than whatever we have chosen as our index of monogamy that contributes to testicle size? If the agreement is even better—so good that the eye can detect no difference between the plotted points and the line, as in panel (d)—we should know that something is wrong; nothing in biology is that reproducible or predictable.

When the agreement between two biological measurements is better than it ought to be, one possible explanation is that although we think we have measured two different things, closer study will reveal that we have actually measured the same thing twice in different ways. However, even this will not do if the agreement is perfect, because it does not allow for our inability to measure any biological quantity without error. In this case, all we are left with is the possibility that we have measured one thing in one way and then done some mathematical transformations to convert it into two different numbers that we fool ourselves into thinking of as two different biological variables.

Human nature being what it is, we are not always as skeptical as we ought to be, and even quite reputable scientists—all of us, probably—are sometimes seduced by results that are too good to be true. In this chapter, I shall use as an object lesson a particular example of results that were too good to be true. Such a chapter may seem to contradict the general thesis of the book, but it does not: we should not be interested in finding optimality wherever we look; we should be interested in finding genuine and demonstrable cases of optimality. So, although I believe that some examples of biochemical optimality are established beyond any doubt, we must always approach any new example with an appropriately critical eye. To show that any given system is optimal, we must define appropriate criteria and we must admit the possibility that we could be wrong. A system is not optimal if its properties are inevitable, that is, if any conceivable system would have the same properties and there is no possibility of changing them during evolution. The number of toes that you have is exactly equal to the number of feet that you have multiplied by the average number of toes on a foot. However, you cannot ascribe this remarkable concordance to natural selection, because it follows inevitably from the way the average is defined.

An example is provided by the phantom phenomenon of *entropy–enthalpy compensation*, which attracted the attention and support of some of the most able enzymologists during the 1970s, has undermined the scientific value of at least one important book on biochemical adaptation, and is not dead even today.

Unfortunately entropy and enthalpy are not familiar concepts to most readers, and their meanings are by no means as obvious, even in general terms, as those of testicle size and monogamy. So I need to begin by explaining these terms, both of them obscure in sound, and one of them quite obscure in meaning as well, but both of them necessary for the discussion. Enthalpy is the less difficult of the two. We will not go very far wrong if we take it as synonymous with heat or energy. If we worked at constant volume (as chemists interested in reactions of gases often do) we could just call it the internal energy and leave it at that. However, biochemical reactions are almost never studied at constant volume, but at constant pressure instead. The very small volume changes that accompany chemical reactions in liquids at constant pressure mean that when a reaction mixture expands, it compresses the rest of the universe slightly, and when it contracts, it allows the rest of the universe to expand slightly. These small changes imply that such reactions have to do work on the rest of the universe or have work done on them by it. Consequently the total energy involved in the process is not quite the same as the change in internal energy in the reaction mixture itself. To allow for the difference, it is convenient to define a separate quantity to measure the energy involved in a process at constant pressure, and this is the enthalpy. As I have said, it will not matter much if you just take it as an obscure word for energy.

Entropy is another matter. Not only is the word obscure, but its meaning is difficult as well. The nineteenth-century thermodynamicist Rudolf Clausius invented the word, quite consciously and deliberately making an obscure choice. He thought that if he chose a word sounding vaguely like "energy" based on Greek roots meaning "transformation" he would have a term that meant the same to everyone whatever their language. As the physicist Leon Cooper drily remarked, he succeeded in this way in finding a word that meant the same to everyone: nothing. This is unfortunate, because it is an essential concept, fundamental for understanding much of engineering and chemistry. For the purposes of this chapter we can think of the entropy of a chemical process as a measure of the probability that it will occur when all energetic requirements have been taken care of. In other words, the probability that any process will occur can be partitioned into energetic and nonenergetic considerations. If we ask what is the probability of throwing a six with a dice, the answer depends both on whether the dice is biased, an energetic question, and on how many possible results there are, a nonenergetic question, which has the answer six in this case.

The calculation of how far a chemical reaction will proceed if left sufficient time can thus be partitioned into an energetic component, measured by the enthalpy change, and a nonenergetic component, measured by the entropy change. This is a problem of *equilibrium*, and the science that deals with it is called *thermodynamics*. This is important for much of life chemistry, because much of life chemistry is just ordinary chemistry; nonetheless, it is of limited value for understanding what is special about life, because this is a question of *rates*, not equilibria. All of the equilibrium positions are exactly the same in living systems as they are in nonliving systems, because a catalyst, whether an enzyme or anything else, has no effect whatsoever on the position of an equilibrium, it only affects how quickly it is reached. All living systems are very far from equilibrium, and without catalysts they would reach equilibrium so slowly that for practical purposes they would not proceed at all. As a result, their states at all moments are determined almost entirely by rates, and hence by enzyme activity, and not by equilibria.

Despite this, thermodynamic ideas are useful for understanding rates because, with some additional assumptions that I shall gloss over, effects on rates can still be expressed in terms of enthalpy and entropy. When a reaction occurs, the substances that react are supposed to pass through a high-energy state called the *transition state*. Because this is a high-energy state, the reactants can only reach it if they have sufficient energy, which means that they must be moving fast enough when they collide, and this in turn depends on the temperature: the hotter a system the faster all of the molecules that compose it are moving, and so the more energetically they collide. This can be measured as an *enthalpy of activation*, and is the reason why nearly all chemical reactions proceed faster as the temperature is raised. However, even if two molecules have enough energy to react when they collide, they do not necessarily do so, because there are questions of orientation that need to be correct as well: these are measured by the *entropy of activation*.

The general idea may be illustrated by reference to a game like tennis in which a ball has to be hit over a net in such a direction that it will land (if not hit by an opponent first) within limits set by the rules of the game. One can imagine a very young person with the potential to become a great player but lacking the physical strength needed to hit the ball over the net. Such a player cannot impart enough energy to the ball to satisfy the enthalpy of activation. For illustrating the entropy contribution there is no need for imagination as one can find suitable players to illustrate the idea on any public tennis court—players like me, for example, who can easily impart enough energy to the ball, but who are very inefficient at sending it in the right direction, so that it has a very low probability of landing in the intended place. Such players impart too much entropy to the ball, and here we say "too much" rather than too little because entropy increases with the degree of randomness.

Estimation of the enthalpy of activation of a chemical reaction is not very difficult because we know that we are making more energy available as we heat a reaction mixture, and if we determine how much faster it gets as more energy becomes available, it is not particularly hard to convert this into a measure of the enthalpy of activation. Experiment shows that many reactions, whether uncatalyzed or catalyzed by enzymes or other catalysts, increase their rates by about a factor of two for each 10°C increase in temperature, for temperatures around the everyday temperatures at which we live. This could mean that many reactions of all kinds have similar enthalpies of activation, which would have important implications about the sort of events that are taking place at the level of the atoms involved. However, it could also just mean that the reactions that behave in this way are the ones that are easiest to study at convenient temperatures, and if so we may be talking about observer bias rather than anything more fundamental about the world.

It is less obvious how we might measure the entropy of activation. As this concerns how fast the reaction could go if all energetic requirements were taken care of, we can expect to get closer and closer to the pure entropy term by studying the rate at higher and higher temperatures, ultimately eliminating enthalpy considerations entirely at infinite temperature. In reality, of course, we cannot produce temperatures of more than a few hundred degrees in an ordinary laboratory, let alone an infinite temperature, and anyway the reactants will boil or burn long before we reach even the modest temperatures we can manage. But if we treat it as a mathematical exercise we can try to estimate what the rate at an infinite temperature will be if we continue the trend observed at ordinary temperatures. Given that ordinary temperatures fall infinitely far short of being infinitely hot, this may seem a calculation of dubious validity, and so it is for enzyme reactions.

Nonetheless, it is not quite as bad as it sounds, because the detailed theory predicts not a straight-line dependence of rate on temperature, but a straight-line dependence of the logarithm of the rate on the reciprocal of the absolute temperature, that is, on the *reciprocal* of the temperature above absolute zero, which is at −273°C. Now even if ordinary temperatures are infinitely far from infinity, their reciprocals are not infinitely far from zero. In the units commonly used, everyday temperatures correspond to reciprocal temperatures of about 3.3, so we might hope that measurements of a rate over a range of, say, two to four in the reciprocal temperature might allow us to estimate what the rate would be at a reciprocal temperature of zero. This is, however, rather a wide range of temperature, from about 25°C below freezing to about 230°C: not impossible for some chemical reactions, but far outside the range in which we could hope to measure an enzyme-catalyzed reaction; for such a reaction we should be lucky to manage 0–40°C, and 0–20°C would be more typical of the ranges one often sees, because attempts to use wider ranges often result in complications (failure of the basic assumptions) that I shall not go into here. A range of 0–20°C is a mere 3.413–3.663 in reciprocal temperature, and the difference between these two, 0.25, is less than one-thirteenth of the difference between 3.4 and zero.

These considerations are illustrated in Figure 7.2, which concentrates on the ranges of temperature involved in ordinary measurements, and the

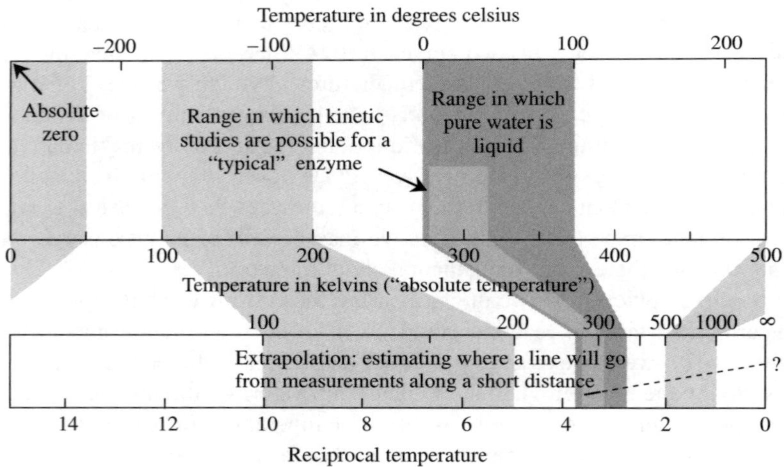

Fig. 7.2 Extrapolating measurements at different temperatures to infinite temperature. The range of temperature over which water is liquid (at ordinary pressures) is 0–100°C, but the scale really starts at absolute zero, −273°C, because nothing can be colder than that. Physicists often prefer to use a scale measured in kelvins (each step of 1 K being the same as a step of 1°C), with absolute zero equal to 0 K. The range of liquid water is then 273–373 K. Even within this limited range, the range in which most enzymes can easily be studied is quite small, typically about 5–40°C, or 278–313 K. Estimating the entropy of a process means estimating what value the property would have at infinite temperature, and to make it easier to do this it is convenient to obtain a reciprocal temperature by dividing 1000 by the temperature in kelvins. Notice that this operation squeezes the range of high temperatures to fit onto the page at the right, while stretching the range of low temperatures to extend off the page at the left. Even on this scale, however, estimating any enzyme property at infinite temperature typically involves continuing an observed trend for more than 10 times its observed range

relationship between temperature and reciprocal temperature. Figure 7.3 illustrates the same ideas, but with emphasis on the specific case of estimating entropy of activation from observations in the range 0–20°C.

It follows, then, that trying to estimate the rate at infinite temperature from observations between 0°C and 20°C is a task for an optimist, and a realistic person will not expect to get more than a very rough idea, at best, of the rate at infinite temperature, and hence of the entropy of activation, from such observations. The process involved is called *extrapolation*, and even a long extrapolation is not always a bad thing to do; sometimes, in fact, there is no choice. If you shoot at a target 150 m away with a rifle of barrel length of around three-quarters of a meter, you are in effect making an extrapolation of 200 times the range of "measurements": the first three-quarters of a meter is fixed by the position of the barrel, the rest is an extrapolation based on a hope that the rifle was pointing in the intended direction when fired, that it is well made enough for the bullet to continue in exactly the same direction when it emerges, and that any effects of wind or gravity have been properly taken into account.

If you are an excellent shot you may well be able to hit a target in these conditions, but most people are not: why are they likely to fail in such an experiment? First, the rifle is likely to be pointing in a slightly different direction from what is intended, and at a distance of 150 m any small error is magnified 200-fold. Second, a poor shot is likely to move the rifle when pressing the trigger. Third, if the rifle is of poor quality or is not intended for use at such a distance, the direction in which the bullet travels may not be precisely determined by the angle at which it is held when it is fired. With most scientific experiments we must think of a low-quality rifle fired by a nervous person with poor vision out of a vehicle moving along a bumpy road on a day with a very strong gusty wind. This is not because most experiments are badly or carelessly carried out, but simply because it is usually impossible to control all the possible sources of error to the point where you can expect to hit a target more than three or four barrel lengths away.

Nonetheless, let us look at what happens if you brush aside all the difficulties and estimate the enzyme activity at infinite temperature anyway. Suppose you extract an enzyme from several fishes that live at different temperatures, from polar seas at around 0°C to tropical lakes at around 40°C, measure the temperature dependence of the activity of each enzyme over a range from 0°C to 20°C (completely outside the living range of some of the fishes, of course), and estimate the enthalpy and entropy of activation of each. If you then draw a graph of the enthalpy of activation as a function of the entropy of activation for all the different fishes, you might expect to be lucky to see any relation at all. You need to remember three things: first, the estimation of the entropy of activation cannot possibly be better than extremely crude; second, the range of temperatures used for the measurements is far too cold for the tropical fishes to live and its upper range far too hot for the antarctic fishes; third, fishes living in African lakes and Antarctic oceans are not likely to be close relatives, and so there are all sorts of reasons why their enzymes might differ beyond the fact that they live at different temperatures.

Despite all of this, when the experiment is done, the correlation is not merely good—but fantastically, amazingly, unbelievably good. Experiments of the general kind I have described have been done by various people, but the particular example that I have in mind comes from a study of the enzyme that powers the muscles of fish (and ourselves, for that matter) done by Ian Johnson and Geoffrey Goldspink. Now, it is perfectly reasonable to suppose that fishes that live at different temperatures may have evolved enzymes that are adapted for the appropriate temperatures, and perfectly reasonable to try to identify the properties of the different enzymes that constitute the adaptations. However, to do this we have to keep within the bounds of what can realistically be measured.

In fact, in this and other similar experiments, all of the points lie on a perfect straight line, so close to it that the eye cannot detect any deviations! Such a perfect result in biology has to be an artifact: there is no way in which it could result from any biological relationship, so it has to be generated by the mathematics. Figure 7.3 shows how this comes about. The experimental points between 0°C and 20°C are too tightly clustered together to give much idea about the best line to be drawn through them. They give a good idea of a central point corresponding to about 10°C that the line has to pass through, but that's about it. We may get a rough idea of the slope to calculate the enthalpy of activation, but any small error in the slope will be translated into a large error in the intercept on the vertical axis; in other words, we cannot determine anything useful about the entropy of activation. In effect, we are trying to estimate two supposedly independent numbers from measurements of one, the activity at 10°C.

Unfortunately, however, many of the people who have done such experiments have not only failed to be amazed but have also invented the fantastical idea of entropy–enthalpy compensation, whereby evolution in its infinite

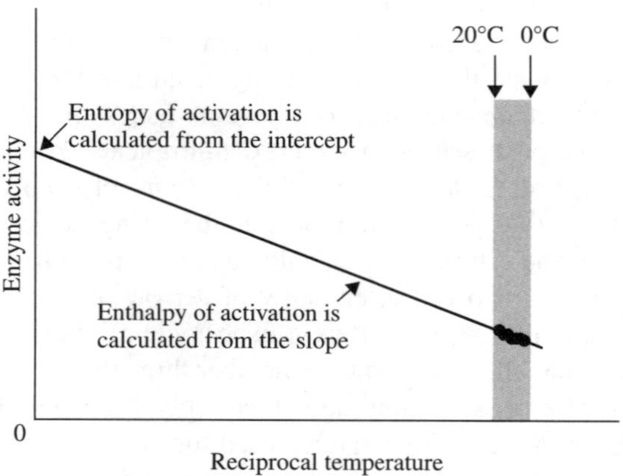

Fig. 7.3 Typical estimation of an entropy of activation. Note that left and right in the horizontal scale are interchanged with respect to Figure 7.2

wisdom has compensated for the changes in enthalpy of activation that result from working at different temperatures, by making changes in the entropy of activation that are just right to produce the same activity at a mystical temperature called the compensation temperature. This, wonder of wonders, often turns out to be within the range over which the measurements were made. Whole branches of the study of biochemical adaptation have been built on this sort of nonsense.

However, before accepting that it is all an artifact of the mathematics, you will need some justification for the claim. Perhaps I can explain it in terms of an analogy involving more concrete things than entropies and infinite temperatures. Living as I do in Marseilles, I have sometimes in summer time had the depressing (but striking) experience of watching the fire-fighting airplanes, or "Canadairs," flying overhead on their way to deal with fires in the hills a kilometer or two from where I live (Figure 7.4). Let us suppose that the smoke is so dense that I cannot actually see where the fire is (fortunately the reality has never been as bad as this!), but I know that the ridge of the hills is 2 km away, and that two Canadairs overhead are 150 m apart. I am interested in knowing both the direction of flight (the "enthalpy") and the location of the fire (the "entropy"). I have no sophisticated measuring instruments, and can estimate the positions of the Canadairs only roughly, but I have maps, graph paper, and mathematical knowledge. So, assuming that they are flying one behind another, I can estimate their direction; continuing the line connecting their two positions at the moment of observation I can estimate where the fire is. Doing this on several occasions during different fires I notice a curious fact: if I make a graph of the direction of flight as a function of the point along the range of hills where I estimate the fire to be, I find a perfect relationship between them, far better than I could reasonably expect in view of the crudeness of the original measurements and the fact that the Canadairs have to fly more than 13 times

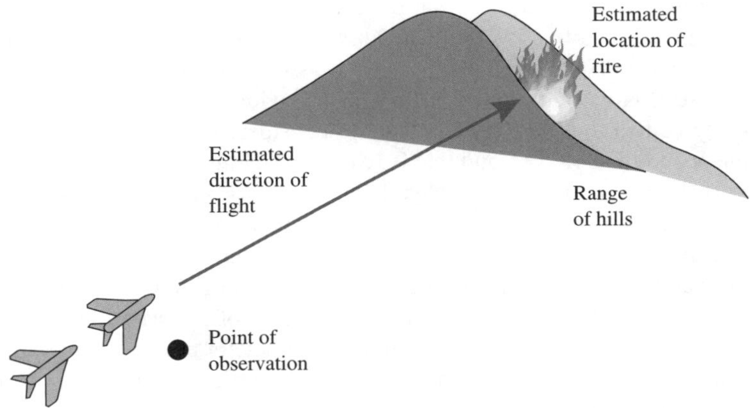

Fig. 7.4 Analogy of entropy estimation. How accurately can one locate a fire by observing airplanes on their way to it?

further than the distance over which I have observed them. Forgetting that both measurements come from exactly the same observations, transformed in two different ways, and that I would not have seen the Canadairs at all unless they flew approximately overhead, I postulate a compensation effect whereby the controller of the fire-fighting team organizes matters in such a way that the direction of fight and location of the fire are correlated to make all of the Canadairs fly over a special point, the "compensation point," which turns out to be very close to the point at which I made my observations. This is a surprisingly accurate analogy to the evidence for entropy–enthalpy compensation.

You might suppose that this was all that needed to be said about entropy–enthalpy compensation, but there is more, because to make a usable theory about biochemical optimization we have to be able to see some biological point in the property that is supposedly being "optimized." What if the phenomenon in question were not just a mathematical artifact? What if there really were a compensation temperature at which the enzymes from fishes living at different temperatures from one another (and different from the compensation temperature) all had the same activity? What would be the advantage to a fish that spent its entire life in tropical waters at 40°C if its muscle enzymes, once removed from the killed fish, proved to have the same catalytic activity when cooled to 18°C as the corresponding warmed enzymes from a quite different fish that spent its life in the near-frozen waters of the Antarctic? Why should such a correspondence have been selected by evolution? I can see no reason, and so even if the phenomenon were based on a much firmer basis of experimental fact than it is, I should be deeply suspicious of it until some plausible advantage to the fishes had been identified.

Actually the story does not end even there. As I have mentioned, studying the temperature dependence of enzymes is often restricted to even narrower ranges of temperatures than the 0–40°C that we might suppose accessible for almost any group of enzymes. This is because estimation of enthalpies and entropies of activation requires that the changing temperature affects only the rate, but in practice this is often true only in a narrow range; for example, a phase change (melting or freezing) in the associated biological membranes may cause a large and abrupt change in behavior at some temperature. In the study of fish muscle enzymes that we considered earlier, such an abrupt change occurred at 18.5°C, and that is why all of the measurements were done below that temperature. This means, however, that if entropy–enthalpy compensation really does represent an evolutionary optimization then we must suppose that certain tropical fishes have "optimized" the behavior of their enzymes at a temperature that is not only well below the temperature at which they spend their entire lives, but on the other side of a discontinuity. It is as if the old-world monkeys had optimized an anatomical feature valuable for defense against piranhas—no use to them in Africa, but it might come in useful if some of their descendants found themselves in Brazil.

I can hardly emphasize too many times that this is not how evolution works: evolution has no foresight; evolution prepares nothing for a hypothetical rainy day; evolution works only in the here and now. Of course, remembering the slowness with which evolution proceeds we should not take "here and now" as too short a period of time. Anything an individual encounters between birth and parenthood is within its present, and as fixation of a gene takes far more than one generation (see Chapter 2) the "here and now" for a species lasts for many generations. That is why individuals retain the detoxification pathways (Chapter10) they can use to cope with poisons that they may not meet in their entire lives. As long as a chemical threat occurs often enough during the long march toward gene fixation, the means to combat it will not get eliminated from the gene pool by natural selection.

However, if an "adaptation" is only useful in some hypothetical future environment it is not an adaptation and will not be selected. Every single change along the chain from primitive to modern must be at least as good as the currently available alternatives, or it will not be selected. This is the famous eye problem—what is the use of half an eye?—that delights creationists so much, because they think that evolutionary theory has no answer to it. Actually evolutionary theory has an overwhelmingly convincing answer to it, but do not just believe it because I say so: read an expert account such as that of Richard Dawkins in *The Blind Watchmaker*.

To guard against similar errors, I have tried in the earlier chapters of this book to give examples of biochemical properties that are not only demonstrably optimized according to some defined criteria, but that also bring some definite biological benefit to their possessor. In discussing the pentose phosphate pathway (Chapter 5), for example, it was not sufficient just to show that the results appeared to satisfy the rules of an arbitrary game; I also had to show that this allowed the organism to grow faster and to waste less of its resources, and thereby compete more effectively with other organisms that had played the game less well.

To end this chapter on a more positive note, I should repeat that searching for adaptations that make particular enzymes best suited to their owners' environments is not by any means a pointless exercise. Many years ago Ernest Baldwin wrote a book that attracted the attention of many biochemists to the interest and importance of comparing the ways different organisms cope with their environments. Since that time, Peter Hochachka and George Somero have devoted much of their careers to this subject and have discovered numerous interesting examples. However, entropy–enthalpy compensation is not one of them.

8
Living on a Knife Edge

It has often been said that "pools must be maintained at their proper levels" or that there is a "normal" level, "excess" of which would either be uneconomic or would upset "the delicate equilibrium" so necessary to "integrate the different metabolic functions." Natural selection has been invoked as being responsible for this amazing feat of juggling. Those who are aware of the forces responsible for coming to a steady state realise, of course, that this is a fanciful delusion. Almost any set of enzymes will generate a steady state with all fluxes in operation. The existence of the vast array of genetic variation shows that there are very many different "delicate equilibria" which are just right. As Mark Twain observed, while marvelling at our amazing adaptation: "Our legs are just long enough to reach the ground."

H. Kacser and J. A. Burns (1973)

Again, hear what an excellent judge of pigs [H. D. Richardson] says: "The legs should be no longer than just to prevent the animal's belly from trailing on the ground. The leg is the least profitable portion of the hog, and we therefore require no more of it than is absolutely necessary for the support of the rest." Let any one compare the wild-boar with any improved breed, and he will see how effectually the legs have been shortened.

Charles Darwin, *The Variation of Animals and Plants under Domestication* (1868)

This chapter deals with another false trail in biochemical optimization, but a quite different one from that of the previous chapter. Here we shall discuss *biochemical regulation*, the general term given to all the effects on enzymes that prevent them from acting when they are not needed, and ensure that they act at the rates that are appropriate for the circumstances. We shall be concerned now with real observations, not fantasy, and the biological advantages will be clear enough; after all, it would be quite inconvenient if our legs were *not* long enough to reach the ground. However, these characteristics are not enough to save an example of optimization from being trivial. If a property, no matter how advantageous, is inevitable, so that we cannot easily imagine matters being otherwise it is not an example of adaptation or optimization, and requires no evolutionary explanation. This has an unfortunate echo of

Dr Pangloss, who thought, at the beginning of Chapter 3, that *because the world was perfect* it could not be otherwise than that spectacles existed to fit over noses, and so on. However, I am not asking you to imagine a perfect world here, but an imperfect one in which not everything is necessarily for the best, but in which some consequences follow inevitably from the starting conditions. As we live in a world where gravity is the most obvious everyday force that affects our lives, we cannot conceive of evolving legs that are too short to reach the ground; the fact that they do reach the ground is thus not an example of an adaptation.

Of course, as Charles Darwin reminds us above, human breeders can certainly select for shorter legs in domesticated pigs, and there is no reason to doubt that natural selection has affected leg length as much as any other character. However, the human leg would require a tremendous amount of shortening before it failed to reach the ground (and even pig breeders have not yet managed to produce legs too short for standing on), so the mere fact of reaching the ground cannot be regarded as an adaptation.

The study of biochemical regulation has many useful things to tell us as long as we are careful to separate the fantasy from the reality. Apart from anything else, a better understanding of the way in which metabolic rates are regulated could have saved the large sums of money that have been wasted on hopeless biotechnological wild goose chases in the past couple of decades. We shall see an example of this in Chapter 10, with great efforts being made to increase the activity of an enzyme called phosphofructokinase in organisms like yeast, in the belief that this would increase the rates of industrially useful processes. However, such beliefs are not based on a real understanding of how biochemical systems behave, and do not produce the benefits expected.

During the 1960s in particular, biochemists learned a great deal about how enzymes are "regulated"; they found that many enzymes did not obey the simplest rate laws possible but instead responded to various kinds of signals that told them whether particular products were appearing in sufficient amounts, or too much or too little. All of this reminded the researchers of the time of feedback control systems designed by engineers, and much of the same terminology was adopted to describe what came to be seen as the perfect design of metabolic systems at the hands of natural selection.

There was much that was good in this, and few would seriously doubt that natural selection is indeed at the heart of the properties discovered. However, there was also a tendency to be overimpressed, and to see living organisms as being far more precariously balanced on a knife edge than they actually are. As Henrik Kacser and Jim Burns pointed out in the passage quoted at the beginning of the chapter, almost any set of enzymes, with almost any reasonable kinetic properties (including an inhibitory effect of each metabolite on the reaction that produces it), will lead to a *steady state* with all processes operating.

This is a state similar to what you can observe in a river in calm conditions: although water is constantly being supplied from above, it is flowing away at exactly the same rate, with the result that not only does the flow rate remain unchanged as long as you continue observing, but also the level of water at the point of observation remains unchanged as well. This example of a river underlines that a steady state does not imply complete inactivity and is perfectly consistent with continuous flow. In a place like Los Angeles where drivers are more disciplined than they are in Marseilles, an aerial film of a highway in good conditions will show a similar picture, with plenty of motion but no accumulation, cars moving away from any point just as fast as they arrive, with all moving at constant speeds. In a natural system, such as a river in calm conditions, the steady state is normally stable, in the sense that if a system is perturbed it will tend to return to the same state as before. Thus we do not have to be especially impressed when we see that mixtures of real enzymes, whether in the cell or in artificial conditions, reach stable steady states without much difficulty.

Normal organisms can support quite considerable amounts of variation in the activities of their enzymes away from "normal" levels. Individuals are far from uniform in their genetic composition, yet nearly all are "normal," in the sense that they can live in a healthy state long enough to become mature enough to leave offspring. The proportion of individuals that reach this level of maturity varies enormously between species: among humans in the western world, for example, the majority of individuals reach childbearing age, and this is also true of some animals living in the wild. But, among most fish, by contrast, the overwhelming majority are eaten by predators long before they reach maturity. In the latter kind of species, the few that escape predators may do so in a few cases because they are better at surviving, but it is far more likely that they are just lucky. In species where comparatively few individuals die before maturity, it is more likely that the ones that die are less healthy, or less good at running away, but even then luck plays a part. But, despite the complicating factors, it is fair to say that the substantial genetic variations between individuals, reflecting substantial variations in the activities of particular enzymes, result in little or no variations in the capacity to survive and leave offspring.

As a specific example, let us consider the human disease *phenylketonuria*, which is caused by the lack of an essential enzyme, phenylalanine 4-monooxygenase. I shall return to this in the next chapter, and for the moment it is sufficient to say that people who lack the enzyme completely have the disease, but people who have only half of the normal amount of enzyme have no related health problems at all: half the normal amount of enzyme appears to be just as good as the full amount. How can this be? If they have half the normal amount of enzyme (as they do, in this and other similar cases), then the reaction the enzyme catalyzes should proceed half as fast, and ought this not have at least some effect? If not, does this not mean that normal individuals have at least twice as much of the enzyme as they need, and that the human species could evolve to become more efficient by decreasing the amount they make, thereby releasing precious resources for other purposes?

This happens to be an example that has been very thoroughly studied in the human species, as phenylketonuria is a serious disease that is easily diagnosed in babies at an age early enough to prevent its symptoms from appearing. However, it is far from being a unique example: the level of almost any enzyme can be decreased by half without any observable effect, and this is the biochemical basis of the genetic phenomenon known as *dominance* that I shall discuss in the next chapter. Here I shall not be concerned with the specific factor of two, but with the more general observation that the concentration of almost any enzyme can be increased by any factor with no observable effect, and can be substantially decreased, sometimes to as little as one-fifth of "normal" activity, with little or no observable effect. How can we explain this without concluding that organisms are hopelessly wasteful, making much more of most enzymes than they actually need?

The answer turns out to be one of mathematical necessity, and explaining it will require some care. There are two biochemical points to consider at the outset. In the first place, most metabolic systems spend a large amount of their time in steady states, and we can go some way toward understanding how biochemical systems behave by restricting the discussion to steady states. In doing this, however, we should keep in mind that it is only a beginning, because many of the most interesting moments in the life of a cell involve transitions from one steady state to another.

The second point is that most metabolic reactions proceed at rates that "try to" vary in proportion to the concentrations of the enzymes that catalyze them. By this I mean that if the enzyme concentration were the only thing that changed, all other concentrations remaining the same, then for most reactions the rate would be proportional to the enzyme concentration. In practice various kinds of complications may prevent a strict proportionality from applying even then, but it is accurate enough for most enzymes to be useful as an approximation. This is easily tested in the laboratory, by studying each enzyme in conditions where no other enzymes are present, and where the concentrations of all the molecules that interact with it are kept constant by the experimenter, and it works fairly well most of the time. The complications that arise in the cell arise because the necessary conditions are *not* fulfilled: other enzymes *are* normally always present, and changing the activity of any of them causes changes in the concentrations of its substrates and products, which affect the rates of other reactions, and so on.

The flow of water in a river corresponds to the chemical flow of metabolites through a series of reactions in a metabolic system; the level of water at a particular point corresponds to the concentration of a particular metabolite. So a metabolic steady state is one in which the chemical flow through each metabolite of the system is constant and the metabolite concentrations are also constant. Taking a river as an example has the complication that the source of the water, involving a large number of ill-defined tributaries, is rather vague. However, if we consider water flowing out of a tank instead, we can see that in the strictest sense a steady state is an illusion: although observing it for a short time may give the impression that the flow rate is constant, it is obvious that as time progresses the amount of water left in the tank must decrease, and with it the pressure driving the water out; the flow rate cannot remain constant, therefore, but must decline.

If we study a chemical reaction on its own, without a mass of other reactions occurring at the same time, we find that it behaves much like water flowing out of a small tank, or out of a child's dam on a beach: the concentrations of any intermediates in the process (corresponding to water levels at different points along the route) initially rise very fast, pass through a maximum and then fall back to zero again; during the same period the flow at any point behaves in much the same way. So although the idea of a chemical steady state arose originally in the context of ordinary reactions it is not really very helpful for analyzing them, because there is no steady state (other than the final state of equilibrium, when nothing is changing), except for an instant when any given intermediate passes through its maximum concentration.

Does this mean, then, that the steady state is just a meaningless irrelevance with no useful application to real systems? In fact no, because in living systems we do find that the flow is virtually unchanging over time and all the way along a pathway. Even if the pathway is branched, the partitioning of the flow at every branchpoint can remain essentially constant and so the flow within any individual unbranched sequence is likewise constant.

If we think for a moment of a single enzyme-catalyzed reaction—not an entire metabolic sequence—we can obtain a steady state if the number of molecules of enzyme is very small compared with the number of molecules of the substrate, the substance to be transformed in the reaction, as we saw in Chapter 1. Even if all of the individual steps within the process are very fast, as they usually are in enzyme-catalyzed reactions, the need for every substrate molecule to encounter an enzyme molecule before it can react will ensure that the concentration of substrate decreases sufficiently slowly to be regarded as constant during the period the process is being observed.

This describes quite well the conditions in a typical steady-state study of an enzyme in the laboratory, as the experimenter can ensure that the enzyme concentration is small enough for a steady state to be produced. We have a problem, however, if we want to generalize it to the cell, as it turns out that most enzyme concentrations in the cell are very much higher than those used in steady-state laboratory experiments, and are often comparable with or sometimes higher than the concentrations of their substrates. Even then, however, we can have a valid steady state if the metabolic system is being supplied from a reservoir of starting materials sufficiently large to be essentially unaffected by the properties of the system we are studying. This is like studying the flow of water through a small stretch of a stream. As long as the water keeps on arriving at the head of the stretch at a constant rate, we do not have to worry about or even know anything about where it comes from. As long as the stream below the lowest point we are studying is capable of removing all the water that arrives, with no danger of becoming blocked and backing up, we do not need to know anything else about it. As long as water arrives at a constant rate at the head and disappears at the same rate at the bottom, we can observe that the flow settles down to the same rate at all points in between, and the levels of water at all points in between settle down to levels exactly sufficient to sustain this rate. It is a matter of everyday observation that that is how streams behave: if there are no changes in the external conditions, such as the sudden arrival of flood water from above, the flows and levels remain unchanged for indefinite periods; in these conditions, we do not observe sudden emptying of intermediate pools with overflowing of others; we do not see water rushing rapidly through a particular stretch at one moment and becoming stagnant a moment later.

Metabolic systems can be analyzed in the same way, as long as we accept the existence of certain "external" properties that are fixed independently of the system under study. As long as we analyze the flow of metabolites between constant reservoirs of starting materials and constant sinks into which the final products flow, and as long as we do not try to explain the constancy of these reservoirs and sinks in terms of the properties of the metabolic processes that connect them, then any metabolic system will settle into a steady state with constant concentrations and flow rates at all points.

In writing the last sentence of the previous paragraph, I first wrote "almost any metabolic system," but then removed the "almost" as unnecessary. Why the moment of doubt? As anyone who has studied the elementary properties of enzymes will know, all enzyme-catalyzed reactions are subject to *saturation*, which means that they have limits that their rates cannot exceed. The same thing happens with transport systems, as illustrated in Figure 8.1. If the number of people wanting to catch the bus is less than the number the bus can

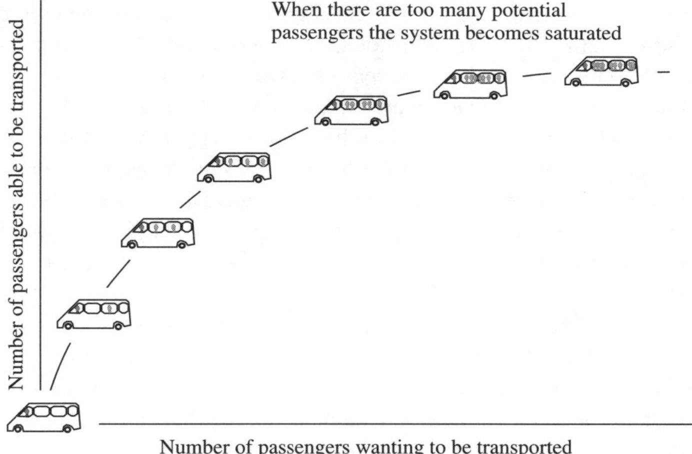

Fig. 8.1 Saturation in a public transport system. When the number of passengers wanting transport is much smaller than the number of places available in the buses, the number transported is the same as the number wanting to be transported. However, as demand increases, and the buses become more crowded, the number of passengers transported falls below the number wanting to be transported

carry, then the number of people transported will be equal to the number wanting to be transported. But the more potential passengers there are the more crowded the buses become, and the more difficult it is for additional people to get into the bus. To visualize this you need to think of a place where the company regulations are not applied very strictly, so the number of passengers is determined by the degree of squeezing they can manage, not by the number of seats in the bus. I once watched in amazement as 12 additional people climbed into a seven-seater minibus in India that was already, in my eyes, full when I started watching.

It follows, therefore, that if an enzyme is being supplied with substrate at a rate that exceeds its saturation limit then it is impossible for it to remove the substrate as fast as it is being supplied, so instead of settling into a steady state the substrate concentration must rise indefinitely. This analysis is valid as far as it goes, but it overlooks two points: first, all enzyme-catalyzed reactions, in common with all chemical reactions, are reversible, even if in some cases the forward direction is so strongly favored that we can never normally detect the reverse process; second, the rate at which any substrate is supplied is never fixed by some external decision but is a consequence of the concentrations of its own starting materials and the properties of the enzymes or other processes that supply it. If these factors are taken into account, then absolutely any metabolic pathway will adjust into a steady state if it is supplied by an unchanging reservoir and flows into an unchanging sink.

As noted by Kacser and Burns in the quotation with which this chapter opened, therefore, the appearance of metabolic steady states is a mathematical necessity that does not require natural selection or any other special mechanism to explain it, any more than we need natural selection or the guiding hand of God to explain why storms do not spontaneously develop in small streams in constant conditions. Once we accept this, we can start to look at some of the properties of metabolic steady states that allow us to understand which characters of metabolism can be acted on by natural selection, and which cannot. These properties form the core of what is now known as *metabolic control analysis*, and were mostly derived by Kacser and Burns in Edinburgh in the early 1970s. Reinhart Heinrich and Tom Rapoport were doing parallel work at the same time in Berlin, and they arrived independently at some of the same conclusions.

Ever since metabolic regulation started to be seriously studied in the 1950s accounts in general textbooks of biochemistry have been dominated by the myth of the *rate-limiting enzyme*. This is the idea that within any metabolic pathway there lurks an enzyme in which all the regulatory properties are concentrated. Change the activity of that enzyme and you will change the properties of the pathway proportionally. For the student of metabolic regulation this is a very appealing idea, because it means that to understand the regulation of a given pathway you can concentrate all the efforts on a single regulatory enzyme, ignoring all the others.

Even in its most extreme form this is not actually impossible: we can indeed imagine a pathway for which the rate is exactly proportional to the activity of one enzyme (at least over a limited range) and completely independent of the properties of all the others. The problem is not that it is impossible but that it is not at all likely, as we shall see, and also that in practice biochemists have often suggested the existence of more than a single rate-limiting enzyme in each pathway. Depending on whom you talk to, you will find any of several glycolytic enzymes described as "the" rate-limiting enzyme for glycolysis. Some textbooks offer different candidates on different pages without any apparent awareness of a contradiction. The reality is that, in most pathways, the flux control is shared (unevenly) among all the enzymes, and we need now to examine why this should be so.

Let us ask the question what will happen to the flux through the pathway if the activity of some enzyme E_1 is increased. In real biochemical engineering projects we are likely to be interested in quite large increases, say 2-fold or 10-fold. However, the analysis is much simpler if we consider very small changes, so here we shall ask what will happen if the enzyme activity is increased just by 1%. If the rate-limiting enzyme idea is right then the answer may be either nothing, if E_1 is not the rate-limiting enzyme, or an increase of 1%, if it is. However, let us not assume the answer at the outset, but say that a 1% increase in the activity of E_1 brings about a C_1 percent increase in flux: if the rate-limiting enzyme idea is right then C_1 is either 0 or 1; if it is not right then it could be anything.

If we now do the same thing with a second enzyme E_2, producing an increase of C_2 percent in flux, the combined effect will be approximately $C_1 + C_2$ percent. We need the qualification "approximately" because we cannot be certain that the effect of changing the activity of E_2 will be the same regardless of whether that of E_1 has changed or not, and because strictly speaking percentages are not additive in this way: if your salary increases by 30% and then increases again by 20% (lucky you) the combined effect will not be an overall increase of 50%, but one of 56%, because the 20% is not 20% of the original salary but 20% of the result of increasing it by 30%, that is, $1.3 \times 1.2 = 1.56$. However, as long as we confine discussion to small changes, the first objection is probably trivial and the second is certainly trivial: an increase of 1% followed by a second increase of 1% produces an overall increase of only slightly more than 2%, close enough to 2% for the difference to be virtually imperceptible.

If there are n enzymes altogether in the system, and if we increase the activity of each of them by 1%, then the overall effect on the flux will be, by an obvious extension of the same argument, $C_1 + C_2 + C_3 + \cdots + C_n$ percent. However, we can evaluate this sum by a different kind of argument. Suppose we increase the rate of every single individual step within a complicated process by exactly 1%, and at the same time decrease the unit of time by exactly 1%, that is, instead of measuring time in minutes we measure it in units of 59.4 s.[1] Let us call this new unit of time a *minnit*. A little reflection should convince you that the new system, with all rates measured in terms of minnits is numerically the same as the old one with rates measured in terms of minutes; anything that previously took 1 min will now take 1 minnit, and as all the rates are thus unchanged so also will be the metabolite concentrations, and as these do not include a unit of time they will be truly unchanged, and not just unchanged by compensating for the real change with a change of unit of measurement.

Now I must confess that when I devised this argument based on changing the unit for measuring time, which is illustrated in Figure 8.2, I thought I had found an easier and more convincing way of explaining the effects of multiple changes in enzyme activities than I had read in previous articles on the subject. However, it is always instructive to find out what others think of supposedly "simple arguments," and when the original version of this chapter was sent by the publishers to several people for their comments, one of them wrote as follows:

> *He says that "a little reflection should convince" me that if I just allow my clock to run a little faster, all will appear the same as before. I am sorry to disappoint the author but quite a lot of reflection has failed on me. I would just love to discuss this with the author and tell him why I can't quite get there, but that's not what happens in a book.*

So, although I really did think that a little reflection would be enough, I was probably wrong, and I shall now approach the same idea from a different direction.

Suppose that we have two identical tanks, each with a tap of identical design, and each with an outlet of identical design, and suppose that we start filling the first tank with water at a constant rate of 10 liters/min, with the outlet left open, and we do nothing with the second one. Initially the water will drain out of the first tank more slowly than it enters, as there will be no head of water to produce the pressure necessary to drive it out as fast as it comes in. However, this will cause the water level to rise, which will generate a pressure, causing the water to drain more quickly. If the tank does not overflow first, the water will rise to a level exactly sufficient to produce a pressure on the outlet just enough to force the water out at 10 liters/min. Let us suppose that this level is equal to 0.2 m (the exact value is not important). The tank will then be in a *steady state*, in which the water is entering and exiting at exactly the same rate of 10 liters/min, and the level inside the tank is exactly constant at 0.2 m.

All this is easy enough to imagine, and most of us are familiar enough with water tanks to believe without checking experimentally that things will happen in the way I say. However, if you are like the anonymous reviewer of this chapter you may still be unconvinced. In this case it is easy to test with an ordinary

[1] Strictly the unit should be 60 s divided by 1.01, or about 59.406 s, but the error that results from ignoring this slight correction is small enough to ignore.

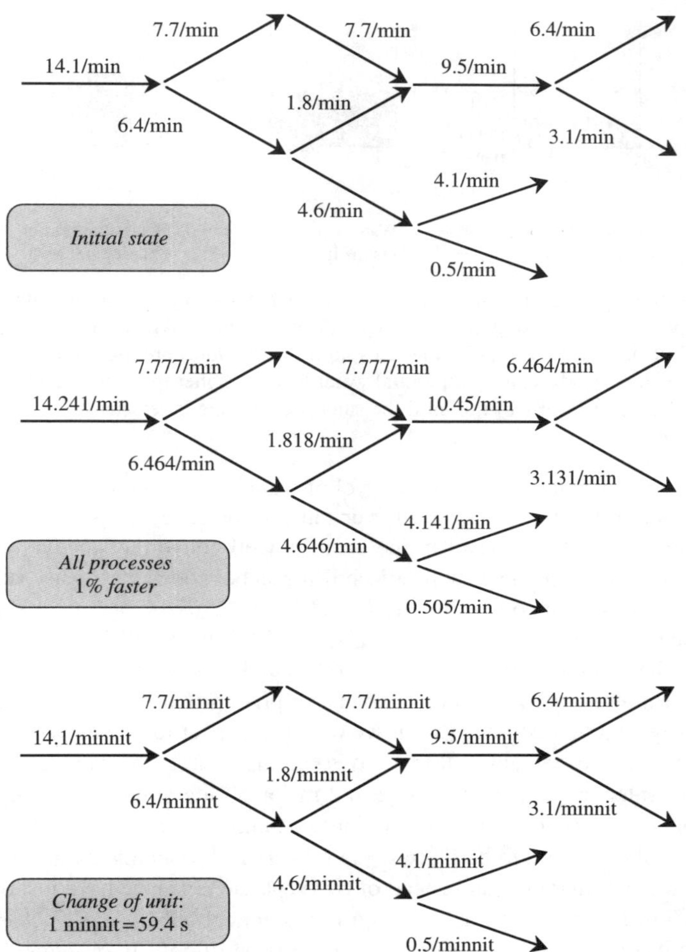

Fig. 8.2 Increased rates and decreased unit of time. A composite process may contain numerous component processes proceeding at different rates. If all of the rates are simultaneously increased by 1%, the new system will not only be physically different from the original one, it will be numerically different as well. However, if the minute is replaced as unit of time by the "minnit," defined as 1% less than a minute, or 59.4 s, the faster system will be found to be numerically identical to the original one, even though it is physically different

basin (Figure 8.3). Leave the tap running at a constant rate (fast, though not fast enough to flood your house by causing an overflow), and do not plug the exit. You will see that the water level will rise, fast at first and then more slowly, and will stop rising when the pressure is enough to force the water out through the plug-hole at the rate it is coming in from the tap. If you repeat the experiment several times you will find that a consistent flow from the tap

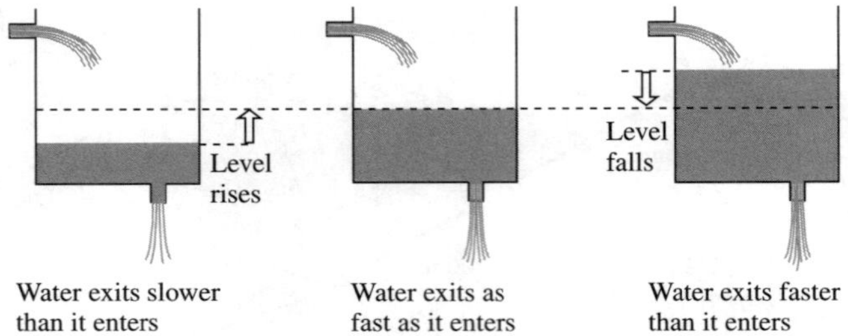

Fig. 8.3 Steady state in a water tank. If water is entering faster than it exits (left) the water level rises, increasing the pressure on the plug-hole, and hence increasing the rate at which it flows through. This continues until the two rates are the same (center), and the tank is in a steady state. If the initial water level is higher than the steady-state level (right), the water level will fall until the same steady state is reached

produces a consistent steady-state level of water in the basin, that higher rates correspond to higher levels, and lower rates to lower levels. Moreover, if you block the exit until the level in the basin is well above the steady-state level and then open it, the level will fall until it reaches the same steady state as it approached from the other direction.

Once you are convinced about this, consider what will happen if you do exactly the same experiment with the other tank. As we have said that this is identical in every way, it would be very surprising if it did not behave in the same way, and that is indeed what we ought to expect (and what you will find if you do the experiment), all the way to giving exactly the same steady-state level of water for a given flow rate, 0.2 m for 10 liters/min, according to our hypothesis. Let us now imagine that our two tanks have a design with a vertical wall that allows them to be placed side by side and touching, as in Figure 8.4. In fact we can imagine that instead of two separate tanks we have just one tank with a thin partition separating it into two identical halves, each with its own tap and its own outlet. However, as long as the partition is in place neither tank "knows" that the other exists, and each will behave as if the other was not there. So with just one tap running we shall have 10 liters flowing per minute and a level of 0.2 m; with both taps running at 10 liters/min the combined rate will be 20 liters/min, and the level in each of the halves will be 0.2 m.

What will now happen if we abruptly remove the partition? Why should anything happen? The levels are the same on the two sides, so there is no reason for any net flow of water from one side to the other; both plug-holes are under the same pressure as they were when the partition was in place, and so they will continue removing water at the same rate. The partition is therefore unnecessary. The consequence of all this is that doubling the capacity of the taps to supply water, and simultaneously doubling the capacity of the plug-holes to remove it, has the effect of doubling the flow without changing the level. However, there is nothing special about a factor of two: if you thought it worthwhile you could do a much more elaborate experiment with, say, 26 tanks

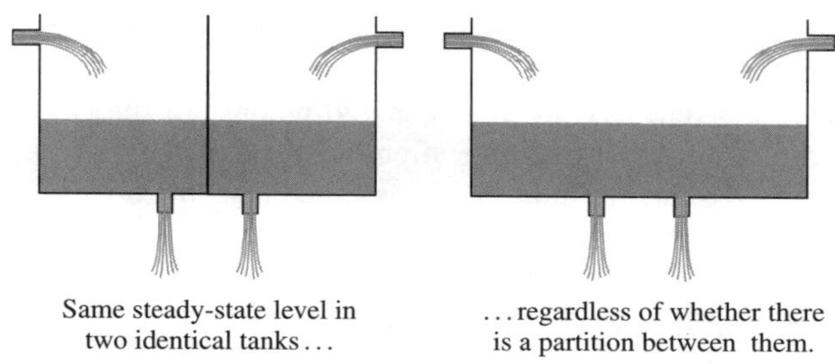

Fig. 8.4 Identical steady states in identical tanks. If two identical tanks are studied with identical input flows and identical plug-holes, they will reach the same steady-state level. Nothing of this will change if the partition separating them is removed, so that they become one tank with two taps and two plug-holes

and could compare the effects of having all 26 in use with that of having just 25 in use: in this case you would find that increasing the capacity of the taps and plug-holes by 4% will increase the flow by 4% and leave the level unchanged. The same result applies to any fractional change you care to test: increasing the capacity of the system by any factor increases the flow by that factor and leaves the water level unchanged.

Here I have avoided any abstract idea like changing the units for measuring time, and I have discussed tanks of water rather than the less-familiar idea of enzymes and metabolites, but the behavior is just the same: if the activity of every enzyme is increased by 1%, then the flux through the pathway (or the flux through each branch of it if there are more than one) will increase by 1% and the concentrations of all the metabolites will be the same as they were before the enzyme activity was increased. So it follows that the total we originally wrote as $C_1 + C_2 + C_3 + \cdots + C_n$ percent can equally well be written as 1%, or, as we do not really need the percent signs, $C_1 + C_2 + C_3 + \cdots + C_n = 1$. This result is called the *summation theorem* or, if we want to be more precise, as there are other summation theorems that define other metabolic variables like metabolite concentrations, the *flux summation theorem*. The individual C values were called *sensitivities* by Kacser and Burns when they first presented the theorem (and *control strengths* by Heinrich and Rapoport), but are now more usually called *control coefficients* or, more precisely, *flux control coefficients*.

As mentioned, there are other summation relationships that can be derived in a similar way. For example, in arriving at the result above, we deduced that a 1% increase in the activity of every enzyme would produce not only a 1% change in flux but also a zero change in the concentration of any metabolite. It follows from this sort of consideration that if we define concentration control coefficients for any particular metabolite, they will sum to zero. These other relationships have, in general, received less attention than the flux summation theorem, which is now regarded as being fundamental to the understanding of metabolic control.

The simplest way of interpreting it is to suppose that there is a finite amount of control over the flux through a pathway, or through any branch of any pathway, and that this control is shared among all the enzymes of the system. There is, however, a complication that needs to be disposed of before taking this idea too far. We have tacitly assumed that flux control coefficients must be positive, but this is not necessarily true. It is already obvious that concentration control coefficients can be negative; otherwise the only way they could add up to zero would be if they were all exactly zero, which is hardly believable (and demonstrably not true, if we analyze the behavior in more rigorous detail than is appropriate here). If we know that concentration control coefficients can be negative, why should we suppose differently for flux control coefficients?

If we return to the analogy of the small stream, it seems obvious that any improvement we might make to the flow at any point—removing a rock, digging a deeper channel, etc.—must improve the flow as a whole (though not necessarily by a significant amount). This is also true in the metabolic system: as long as we are dealing with an unbranched chain of reactions, any improvement at one step will be an improvement overall (though again, not necessarily by a significant amount). However, once we admit branches to the pathway matters become more complicated. Suppose that instead of being straight, our stream divides into two channels as it passes an island. It now becomes easy to believe (and is true) that improving the flow through one of the channels will tend to decrease it in the other. The same is true in a branched metabolic pathway: increasing the activity of an enzyme in one branch will tend to increase the flux in that branch (the relevant flux control coefficient will be positive) but can decrease it in the other (the corresponding flux control coefficient now being negative).

This clearly complicates the interpretation of the summation theorem, as the idea of sharing is less clear if some of the shares can be negative. We can easily imagine having several individual shares of one if there are enough negative ones elsewhere to give a total of one. Easy to imagine, perhaps, but not so easy to design such cases in the computer or to find examples of them in real studies of metabolism. In the computer, it is quite easy to define conditions where some of the flux control coefficients have small negative values, but much more difficult to create conditions that will produce large negative ones. We are forced to replace our nice clean mathematical statement that control must be shared because all the values must add up to exactly one by a less-tidy statement: although negative coefficients are possible, they are not particularly common and when they occur they are usually small, so the positive coefficients add up to a number that is not much bigger than one. In Chapter 10 we shall see that the liver enzyme hexokinase D is an exception to this statement and, more important, we shall be able to understand the reasons why it should be an exception.

Although messier than we might like, this statement still provides us with a useful basis for understanding metabolic control. It still allows us to say that control is approximately shared among all the enzymes in a system, that the average flux control coefficient will be about one divided by the number of enzymes in the system, and that unless there are special mechanisms that ensure that all the control is concentrated in one enzyme there is no particular reason to expect the flux to be proportional to the activity of any given enzyme.

For detailed analysis, whether theoretical or experimental, we usually take "the system" to consist of a short pathway of three or four enzymes connecting two reservoirs. However, it does not have to be as simple as that. At least approximately we can take the whole organism to be "the system," with one reservoir consisting of food and oxygen, the other consisting of waste products. Large animals like humans neither feed nor excrete continuously, of course, but they do so often enough for there to be long periods when the bloodstream is continuously receiving the digested intake from the last meal, the lungs are continuously taking in oxygen and removing carbon dioxide, and the bladder and bowel are being continuously filled with other waste products. Because of variations in the level of activity—sleeping, eating, walking, running, etc.—the metabolic rate is not constant either, but neither does it vary so enormously that we cannot reasonably ask how its average value might vary in response to the activity of one enzyme. In this case, therefore, the system contains many thousands of different enzymes, and the average flux control coefficient is much less than 0.001, effectively zero when we take account of the difficulty of measuring it accurately. It follows that if we choose any enzyme at random and vary its activity by a few percent then we will in all likelihood detect no change whatsoever in the metabolic rate of the organism.

It is only one step from here to start thinking that if changing the activity of an enzyme by a few percent has no detectable effect, then the organism has more of that enzyme than it needs: the enzyme is "in excess." For many enzymes, we might well find that we could decrease their activities by as much as one-half without seeing much effect. Is this not wasteful? Could not the organism become more "efficient" by making only half the amount of each such enzyme? But we need to think more deeply before reaching any such conclusion. To simplify matters, suppose we are dealing with an enzyme with a flux control coefficient of exactly 0.001 and normally produced in sufficient quantities to constitute 0.1% of the total protein in the organism. Let us now consider how much "wasted effort" we could "save" by synthesizing 1% less of it. (As when we introduced the concept of control coefficients, we would really be interested in making a much larger saving than this, but the analysis is easier if we consider small changes.) Doing this would save 0.001% (1% of 0.1%) of the protein-synthesizing investment of the organism, and it would lower its metabolic output by 0.001%. This latter figure is certainly negligible, as we should never be able to detect so slight a diminution of output. So apparently we have something for nothing: a saving in investment for no detectable cost. Or have we? The problem is that the "saving" is just as negligible as the "cost," as it has exactly the same value of 0.001% of the total. The organism would no more notice a saving of 0.001% of its investment than it would notice a 0.001% loss of output.

As soon as we start to be more greedy, saving substantial amounts of the "unnecessary" enzyme activity, and decreasing the activities of numerous enzymes rather than just one, the savings become much more significant, but so, unfortunately, do the costs. On the simple proportional model we have been adopting, the costs would just increase in direct proportion to the savings: if we were greedy enough to try to manage with one-third less protein synthesis, for example, we should end up with one-third less output. Even with this model it is obvious that the "saving" is not working out as we hoped: we are paying less and receiving less in return. In reality, however, it would be worse than this, because flux control coefficients do not remain constant when conditions change: almost invariably a flux control coefficient increases when the activity of the enzyme it refers to decreases. It is not difficult to see in general terms why this should be so: no matter how unimportant an enzyme may be, it will eventually become the weakest link in the chain if we decrease its activity enough; in other words as its activity approaches zero, its flux control coefficient for the flux through the particular branch where it is located must approach unity. It follows, therefore, that if an enzyme activity decreases by more than a few percent, the cost in terms of lost output will be larger than the saving.

More generally, it follows from this discussion that the title of this chapter is a nonsense, and that neither we nor any other organisms are living on a knife edge where any departure from the delicate balance of enzyme activities that natural selection has achieved during many millions of years will be disastrous. On the contrary, most metabolic steady states are remarkably robust, and can tolerate quite large changes in the activities of many different enzymes. This has many consequences, some of which we shall be examining later in this book (Chapters 9 and 10), but first we must recognize and explain one respect in which metabolic systems are quite different from flowing systems in the nonliving world, like rivers.

If you examine any lake in the world—I know of no exceptions—you will notice that although there will normally be numerous points where water enters, there will be a maximum of one point at which water exits (Figure 8.5). There may be no exits at all (other than evaporation), as with the Caspian or the Dead Sea, but usually there is exactly one. Let us take the Lake of Geneva as an example. Normally it is said that the Rhône flows into it near Montreux and that it flows out at Geneva. True enough, except that the Rhône is not the only river that flows in: there are also the Hermance, the Dranse, the Vevey, the Morge, and so on, at least a dozen large enough to be marked in road atlases, not to mention numerous other streams too small to be noted on anything but a large-scale map. If we follow the Rhône further as it flows from Geneva to the sea, we find numerous other streams that join it, some of them large rivers like the Saône or the Isère, but it does not divide permanently into two until the Little Rhône separates at Arles. By there, however, we are well into the

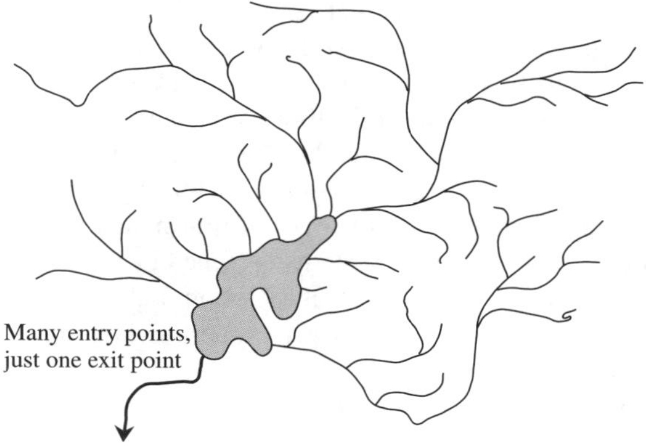

Fig. 8.5 Unregulated flow. If there is no interference by humans or beavers, a typical lake has many inputs and exactly one exit. (Lakes with a very high rate of evaporation, such as the Dead Sea, may have no normal exits at all.)

region of human control of the flow, so we cannot attribute it to natural forces. Nonetheless, it is not uncommon for a river to divide into multiple channels as it approaches the sea across a nearly flat flood plane, but this is hardly an exception to the rule that rivers unite but do not divide, because once a river reaches its delta, the flow is no longer unidirectional.

The behavior of the Rhône is in no way exceptional. To take a North American example, the Mississippi is joined by the Missouri, the Ohio, and numerous smaller rivers along the route to the Gulf of Mexico, but it does not divide until it reaches its delta below New Orleans. These examples illustrate, therefore, how flows evolve if left to their own devices without the intervention of selection, whether natural or unnatural.

Controlled flows are very different. If we look at how traffic flows in a road network we find that it divides as often as it unites, and although there may exist major routes onto which the minor ones converge, the points of convergence are also points of divergence. The circulation of blood in the animal body is very similar. Even if there are no two-way roads as such, there are separate routes to allow flow in the two directions, and the many points of convergence to the major arteries and veins are matched by a correspondingly large number of points of divergence.

Even though all metabolic reactions are driven by thermodynamic imbalances, just like any other chemical reactions, and in that sense like the gravitational forces that drive the flow of water in rivers, their arrangement nonetheless resembles that of a man-made network like a road system far more than it resembles the drainage network of a river valley. Even though there is a general direction to the entire process—roughly speaking we can say that food plus oxygen is irreversibly converted into carbon dioxide plus water—there are just as many points of divergence as there are points of convergence. This is not how flow systems behave if left to their own devices, however, and there is no mathematical analysis that leads us to expect it. On the contrary, any simple mathematical analysis would suggest the opposite.

In contrast to the first half of this chapter, therefore, where we saw that some properties of metabolism sometimes mistakenly attributed to natural selection are in reality just the inevitable consequence of mathematical necessity, here we have properties that certainly call for explanation, because they are metabolic properties that do not follow from mathematical necessity but are much more similar to the way systems designed by engineers behave. This explanation can only come from natural selection, but it is not enough to say that, because we need to recognize what properties have been selected that allow such behavior.

Circulating water remains relevant, but now we must look not at river basins but at unnatural networks like irrigation systems. Although in advanced economies these may involve pumps and external power sources, in the simplest cases the driving force is exactly the same as it is for a river, namely gravity, so that the water must always flow downhill. Nonetheless by arranging the channels so that they are almost level but separated from the reservoirs and sinks by gates, it is possible to make the water flow along whichever channel and in whichever direction is required at a particular moment. We shall see shortly what correspond to gates in metabolic systems, but first we must pause to ask whether they rely solely on "gravity," that is, thermodynamic forces, or whether they incorporate anything resembling pumps.

The answer to this is largely just a question of definition. At one level we can certainly say that every metabolic reaction, at all times and in all circumstances, proceeds in the direction ordained by thermodynamic considerations. True though this is, it requires us to think of reactions in terms of all their components, that is, when we consider a reaction we must always consider the complete reaction, explicitly taking account of all of its substrates and all of its products. Often this is not very convenient, however: when we consider the conversion of glucose to glucose 6-phosphate, for example, we often prefer to ignore the other half of the reaction. In converting glucose to glucose 6-phosphate the phospho group has to come from somewhere, and the hydrogen atom that it replaces has to go somewhere.

These requirements can be satisfied by "coupling" the half-reaction that converts glucose to glucose 6-phosphate to some other half-reaction, as illustrated in Figure 8.6, but we are not restricted to a unique possibility for this second half-reaction. It could be coupled to the conversion of ATP to ADP, as we discussed in Chapter 1, in which case it will proceed forwards if the concentrations of all four components are equal; it could be coupled to the conversion of the inorganic phosphate ion to water,[2] in which case it will proceed backwards; or it could be coupled to phosphorylation of a sugar similar to glucose, such as fructose, in which case it will not proceed very far from equality of the four concentrations.

[2] Yes, I know this sounds absurd. I will explain what I mean in a moment.

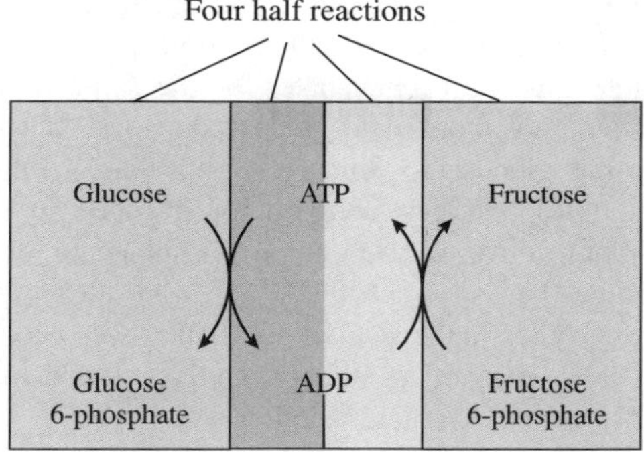

Fig. 8.6 Half-reactions. In many cases, a biochemical reaction can be regarded as the sum of two half-reactions, neither of which could proceed by itself, because a half-reaction does not respect the laws of chemistry. For example, no reaction could convert glucose into glucose 6-phosphate without anything else happening, because glucose 6-phosphate contains (among other atoms) a phosphorus atom not available from glucose. The classification into half-reactions is useful because some half-reactions, such as conversion of ATP to ADP, occur in many different reactions

The first two of these three choices are of great importance in metabolism. Coupling to ATP is catalyzed by the enzyme hexokinase, and when hexokinase is active the traffic through glycolysis, one of the principal highways of metabolism, proceeds almost irreversibly from glucose to smaller molecules. Coupling to inorganic phosphate is catalyzed by glucose 6-phosphatase, and when this enzyme is active the traffic is in the opposite direction. We can regard the half-reactions with ATP and with inorganic phosphate as pumps that decide which direction the half-reaction from glucose to glucose 6-phosphate will go, or we can say that we are talking about two distinct complete reactions, each of which proceeds in the direction ordained by thermodynamics. However we like to express it, the direction of the glucose phosphorylation reaction is determined by which of the two enzymes is active: there is always enough water available to ensure that glucose 6-phosphate can be converted virtually completely to glucose and inorganic phosphate if glucose 6-phosphatase is active, and there is normally (i.e. in a healthy cell) enough ATP available to ensure that glucose will be converted very largely to glucose 6-phosphate if hexokinase is active.

You may have noted an apparent absurdity a couple of paragraphs ago, when I talked about inorganic phosphate being converted to water. What can this possibly mean, given that water contains no phosphorus atom, and the phosphate ion (at least in the triply charged form commonly drawn in textbooks) contains no hydrogen? Surely I am not proposing that living systems can realize the ancient dream of the alchemists, the transmutation of one element into another? No, I am simply underlining that when we talk about a half-reaction the very term implies the existence of another half-reaction that takes care of the necessary bookkeeping that the first one apparently ignores.

The objection that conversion of inorganic phosphate into water is impossible applies with equal force to any other half-reaction: it is no less impossible to convert glucose into glucose 6-phosphate, for the same reason that glucose 6-phosphate contains phosphorus and glucose does not. If we are allowed to talk about converting glucose to glucose 6-phosphate (something that biochemists do all the time) then there is no obvious reason why we should not be allowed to talk about converting inorganic phosphate to water. If doing so sounds rather strange then so much the better, as it alerts us to the dangers in talking about any other half-reactions as if they were complete reactions. The point is that it *should* sound strange, and so should talking about any half-reaction as if it were a complete reaction.

Using gates to control the flow of water in irrigation systems is in principle quite simple, and has proved to be reasonably easy in practice as well, as many societies throughout history have succeeded in designing such systems and managing them effectively. The corresponding metabolic problem is much more formidable. Physical barriers such as cell walls and membranes are used in metabolism to prevent the free transfer of metabolites to places where they are not wanted, but the metabolic network is so vast and complicated that there are many circumstances where the chemical flow needs to be controlled without the use of physical barriers, and this can only be done by varying the catalytic activities of enzymes. Unlike a physical door that can be opened to its fullest extent or completely closed with expenditure of very little energy, varying the activity of an enzyme over a wide range is very difficult.

The most effective way of decreasing an enzyme activity to zero is to break the enzyme down to its aminoacid components, resynthesizing it when it is needed. However, this is too slow for many purposes—when a rabbit sees the fox approaching, it needs to start converting its chemical stores into running energy immediately. There is no question of synthesizing new enzyme molecules to achieve a different metabolic state; it has to make do with changes that can be effectuated instantaneously, and this means using inhibitors and activators. Unfortunately (from the point of view of regulatory design) the small molecules that inhibit or activate enzymes do so by binding to them according to the ordinary laws of chemistry, which put severe limits on the amount of change that can be brought about by reasonable changes in the concentration of inhibitor or activator.

Roughly speaking, unless the enzyme has special properties, a change from 90% to 10% of its full activity requires almost a hundredfold change in the concentration of a small molecule, whether it is its substrate, an inhibitor or an activator that modulates it. This is exactly the opposite of what any reasonable designer would want. When we use a light switch, for example, we expect a tiny input of mechanical energy into the switch to take us all the way from completely off to completely on, or vice versa. Ideally, we should like a small change in the concentration of a signal molecule, say a change of a few percent, to switch a pathway on or off, but this is not at all what we get from the usual properties of enzymes and small molecules. A hundredfold change in concentration is a very large change, and spanning the range from 10% to 90% activity is a rather modest interpretation of what we usually mean by off and on. Matters are made still worse by the considerations we discussed in the first half of this chapter: very few enzymes have anything approaching complete control of the flux through the pathway where they find themselves, and so even if an inhibitor succeeds in bringing a particular enzyme from 90% to 10% of full activity, the net effect on the flux through the pathway will be less than that!

This is rather depressing, but there are a number of partial ways around the problem that are available to living organisms. First, metabolic signals can act at several different sites: there is no requirement that an inhibitor must inhibit just one enzyme and do nothing else; in reality it can inhibit certain enzymes and activate others and, provided the sites of action are carefully selected, all of the effects can point in the same direction. (The use of the word "selected" here is, of course, a deliberate evocation of natural selection: it is quite difficult in this field to avoid anthropomorphic or teleological language that implies the existence of a conscious designer, and the circumlocutions needed to avoid such language are not worthwhile as long as everyone realizes that the long-term effect of natural selection is design without a designer.) Second, more than one small molecule can act on the same enzyme, and even though the effects are not strictly additive they are certainly to some degree cumulative: it is more effective to inhibit an enzyme and decrease the concentration of its activator simultaneously than it is to do just one of them. Third, the 100-fold change in concentration for a 10–90% change in activity that I referred to assumed that the enzyme had "ordinary" properties, but this is not necessary. Enzymes can have structures that allow them to respond more sensitively to changes in concentration, and it appears reasonably easy to decrease the 100-fold range to a range of around 5-fold by using the property of cooperativity that I shall discuss in Chapter 10. Reasonably easy, certainly, but not trivially easy, as most enzymes do not have this property. Presumably there is a price to be paid for it: the cooperative enzyme may be less effective as a catalyst than its non-cooperative analog, or it may require a more delicate and easily damaged structure, etc. Moreover, the fact that extremely few enzymes take cooperativity much further than this—the fivefold range of concentration for a 10–90% range of activity—even though it would appear to be extremely useful for regulating fluxes if they did, suggests that it is not a particularly easy property to design. It may be just ignorance on my part, but I am not aware that anyone has succeeded in producing cooperativity in an artificial catalyst.

The fourth way of getting around the difficulty may seem less general, but it is important because so many metabolic reactions involve ATP, and because the status of ATP as the cell currency means that quite small changes in ATP concentration can provide the cell with important clues to changes in metabolic circumstances. I mentioned in Chapter 1 that in addition to ATP and ADP there was a third member of the adenine nucleotide family, AMP, with just one phospho group. However, other than mentioning that it participated in fewer metabolic reactions than the other two, I was rather vague about its function. In fact, AMP is important in part *because* it participates in few reactions. Because of that, there are few reactions to be perturbed if large changes in its concentration occur. Moreover, its concentration is typically much smaller than those of ATP and ADP, and an equilibrium among the three is maintained in many tissues by an enzyme called adenylate kinase or myokinase. At first sight this is rather a pointless reaction: what is the value of maintaining ATP and ADP in equilibrium with a third molecule that is present in small amounts and not used for anything much? In fact, it is only true to say that it is not used for anything much if we take this to mean that it rarely acts as the substrate or product of a reaction, but there are other things we can use a metabolite for than just using it as the starting material for making something else. In particular, we can use it as a *signal*.

The equilibrium constant for the myokinase reaction is not far from one: this means that at equilibrium we can calculate the concentration of AMP roughly by multiplying the ADP concentration by itself and dividing by the ATP concentration. For example, at ATP and ADP concentrations of 10 and 1 mmol/liter, respectively (chosen as simple numbers to calculate with, but not too grossly far from typical concentrations in the cell) we should estimate an AMP concentration of one-tenth of 1 mmol/liter. Suppose now that the cellular demand for ATP has become so great that its concentration has fallen to 9 mmol/liter and that of ADP has risen to 2 mmol/liter. The concentration of AMP at equilibrium is now four-ninths of 1 mmol/liter. To be more rigorous we should have to take account not only of the fact that the equilibrium constant is not exactly one, but also that the extra AMP has to come from a decrease in the total of ATP and ADP. However, correcting for these points would not make enough difference for it to be worthwhile being more complicated here.

Notice what this calculation means. We have found that a 10% decrease in the ATP concentration has been amplified into a 4.4-fold change in the AMP concentration. The fact that the first three letters of "amplified" are AMP is just a coincidence, as AMP was given this symbol long before its amplifying function was understood, but it is a very happy coincidence that makes it easy to remember one of the main metabolic functions of AMP. As AMP is not needed for its own sake in many metabolic reactions, a fourfold change in its concentration is unlikely to produce any problems. Instead, it can be used as a signal to enzymes that would "like" to be able to respond to a small change in ATP concentration. Designing an enzyme to recognize a fourfold change in AMP concentration is much easier than designing one to respond to a 10% change in ATP concentration. So this is what many enzymes do: wherever logic tells us that it would be useful for regulation for an enzyme to respond to small changes in the concentration of ATP we often find instead (or in addition) that it responds in the opposite direction to changes in the concentration of AMP; enzymes that "ought" to be activated by ATP are often inhibited by AMP; enzymes that "ought" to be inhibited by ATP are often activated by AMP.

The conclusion from all of this is that we do *not* live on a knife edge. We do not need natural selection to explain the observation that much of metabolism can be represented simply as a set of pools of major metabolites at approximately constant concentrations, with chemical flows between them that proceed at rates that over short timescales vary little or not at all. Systems of enzymes catalyzing diverse sets of reactions readily achieve steady states because that is almost an automatic property of such systems. Assigning kinetic properties haphazardly to all the enzymes in a system normally does not produce any exotic properties for the whole system: as Kacser and Burns remarked three decades ago, "almost any set of enzymes will generate a steady state with all fluxes in operation," with intermediate pools at their "proper levels," and so on. So, desirable as these properties may be, we have no need to invoke natural selection to explain them.

9
Brown Eyes and Blue

Biochemically dominance must be determined by a frightfully complex, and perhaps equally delicate, series of reactions.

R. A. Fisher, Letter to C. G. Darwin, July 16, 1930

You may be familiar with Richard Dawkins's books *The Selfish Gene* and *The Blind Watchmaker*, and may perhaps have learned most of what you know about evolution, particularly the evolution of behavior, from these excellent sources. His second book, *The Extended Phenotype*, is less well known, and that is a pity, because in many ways it is his finest achievement. Nonetheless, even the most lucid books contain some obscure passages, and you could well have been puzzled by a couple of pages of *The Extended Phenotype* that deal will the theory of modifier genes, which R. A. Fisher proposed in 1930 to explain why the phenotypes of some genes are dominant whereas others are recessive, "phenotype" being a technical term for the specific set of observable characteristics that indicate the presence of a particular variant of a gene.

In fairness to Dawkins, it must be admitted that these pages are not nearly as obscure and difficult as Fisher's own account of his theory, which occupies a chapter of his book *The Genetical Theory of Natural Selection*. Fred Hoyle, a theoretical astronomer whom I shall discuss in more detail in Chapter 11, recommended this book for its "brilliant obscurity," adding that "after two or three months of investigation it will be found possible to understand some of Fisher's sentences." Even William Hamilton, one of the greatest evolutionary biologists of modern times and a great admirer of Fisher's book, found it heavy going: "Most chapters took me weeks, some months," he wrote.

It may well be true, as Dawkins asserts, that by 1958 the modifier theory was so well accepted, along with Fisher's view that dominance must be an evolved property because it has selective advantage, that he felt no need to justify it when he wrote the second edition of *The Genetical Theory of Natural Selection;* indeed, the theory was still well accepted in 1981, when *The Extended Phenotype* was written. Coincidentally, however, 1981 was the year in which the matter was clarified once and for all by a landmark paper by Henrik Kacser and Jim Burns, and, as we shall see, their explanation differs completely from Fisher's, being instead much closer to the point of view of Sewall Wright, Fisher's great opponent over many years.

Why is there a problem at all, and why did it take well over a century from the time when Gregor Mendel first described dominance for it to be properly understood? Moreover, why were Fisher, Wright, and J. B. S. Haldane, some of the greatest minds to have influenced the course of genetics, unable to agree about it, and, in the cases of Fisher and Haldane, unable even to come close to the right answer to what seems in retrospect quite a straightforward question?

Before trying to answer these questions, we should remind ourselves what it means for the phenotype of a gene to be dominant or recessive. Mendel studied seven different "characters"—green or yellow seeds, wrinkly or smooth seeds, tall plants or short, and so on, but one is enough to illustrate the idea. First we must insist on starting with true-breeding strains, which means that if we cross two plants with green seeds we can guarantee to get another plant with green seeds, whereas if we cross two plants with yellow seeds we can guarantee to get another plant with yellow seeds. So far so good, but what do we expect if we cross a plant with green seeds with one with yellow seeds? Naively, perhaps, we may expect greenish-yellow seeds, but that is not what Mendel found: he found that in the first generation of crosses all the offspring had green seeds. Even more strikingly, if these first-generation offspring plants were crossed with one another, three-quarters of their offspring had green seeds, but one-quarter had yellow. Further analysis would show that the descendants with yellow seeds were like their yellow ancestors—true-breeding if crossed with other yellow-seed plants. Of the three-quarters of descendants with green seeds, one-third were true-breeding and the other two-thirds were like the first-generation crosses.

What all this means is that crossing two individuals is not just a matter of mixing their characters in the same way as you might mix the contents of two cans of paint. Instead Mendel's result shows clearly that inheritance is *particulate*, the particles corresponding to the entities that we now call genes. The simplest interpretation is that a pea plant has two copies of the gene that determines seed color: in the first-generation cross from true-breeding parents with green and yellow seeds, there is a gene for the green phenotype from the green parent and a gene for the yellow phenotype from the yellow parent, but the green phenotype is dominant over the yellow, so that whether a plant has two genes for green color or one, its seeds are green. In Mendel's time, and until the late 1920s, this was just an observation, and no one tried to explain why some phenotypes were dominant whereas others, the yellow phenotype in this case, were recessive.

Few things in biology are simple, and even when it seems possible to make some tidy generalizations, it does not usually take long for some tiresome exceptions to accumulate. So, before proceeding, I should perhaps make it clear that not all genes behave as Mendel observed. First, not all organisms have their genes in pairs, that is, not all are *diploid*. Some, like bacteria, are *haploid* and have only one copy of each gene; for them the concept of dominance has no meaning. Others, such as certain plants, including ones like wheat that are intensively cultivated, are *polyploid*, and have more than two copies; for them dominance exists but its analysis is more complicated. A few insects, such as bees, are *haplodiploid*, which means that males are haploid and females are diploid: this has fascinating consequences for their behavior, but I must resist the temptation to discuss them here as they have nothing to do with the main themes of this book.

Some organisms, such as the alga *Chlamydomonas reinhardtii*, exist mainly as haploids but occasionally pass through a diploid state: this characteristic is highly relevant to this chapter, and I shall return to it later, but for the moment I shall set it aside as an unwanted complication. Most of the organisms likely to interest us, including humans and nearly all other animals, and peas, and most other green plants, are diploid, and for them the principles of Mendelian inheritance apply to all simple traits.

I hope you noticed the weasel-word "simple" that has slipped in here. What is a simple trait? Obviously, one to which Mendelian inheritance applies! Fortunately, however, we do not have to tolerate such a sloppy circular argument, because the more understanding we gain about what genes actually code for the more we can understand that even if one gene does affect just one protein (to a first approximation: I shall not worry about exceptions here), more than one protein may be involved in producing an observable character, and if so then more than one gene must be involved as well. It is worth spending a moment to reflect on this, because Mendelian inheritance applies very well to a great many cases even if at first sight it may seem to contradict our everyday experience. When this happens, we may be tempted to reject a whole subject because it appears to lead to nonsense, though with more careful explanation it might make good sense.

The classic example used in elementary textbooks to illustrate the application of Mendelian inheritance to humans is eye color: the phenotype of brown eyes, it is said, is dominant, whereas that of blue eyes is recessive. True enough, as long as we confine our attention to people whose eyes are bright blue or dark brown; but when I look into a mirror the eyes that gaze back at me are neither blue nor brown, but the sort of vague greenish-brown color commonly called "hazel," and I cannot easily relate what the more oversimplified textbooks say to my own experience. Unfortunately many nongeneticists, including some of the sillier leaders of political opinion, are more interested in skin color than eye color, and more interested in both than they are in breeding peas, and when it comes to skin color the effects of crosses are much more like mixing paint than elementary notions of Mendelian inheritance would lead you to expect.

Does this mean that Mendelian inheritance is a myth, or something that applies well enough to peas but not to species that we care about? No, it means that most of the traits that we can readily observe without instruments, like eye or skin color, are not simple traits. In terms of skin color, it means that there is not just one pigment produced by just one enzyme coded for by just one gene. Moreover, skin color is also affected by environmental, nongenetic conditions, most obviously by exposure to sunlight. Even if we confine attention to purely genetic considerations, several different genes are involved in skin color that interact in ways that are not simple and not fully understood. Moreover, they are not passed on from parent to child in a block, but separately. All of this means that skin color is not a good trait to consider if we are interested in finding a clear illustration of the principles of inheritance. Unfortunately the same complications apply to some of the most important metabolic diseases in humans: most forms of diabetes, for example, are not determined by single genes but result from the combined effects of several genes. For this reason it is only quite recently that much progress has been made in understanding the genetics of such diseases.

It all works much better if we consider some simpler cases that are less obvious in ordinary life but still have greater importance for medicine than skin or eye color. Take, for example, the disease phenylketonuria, which I mentioned briefly in Chapter 8. If left untreated it produces severe mental retardation and, in many cases, death before the age of 25. It is caused by an incapacity to convert the aminoacid phenylalanine into another aminoacid, tyrosine. It is not, however, a deficiency disease, because its harmful effects are not caused by a shortage of tyrosine, and cannot be avoided by adding tyrosine to the diet. Instead they are caused by the toxic effects on the brain of a substance called phenylpyruvate, which the body produces in its efforts to remove the excess of phenylalanine. The name of the disease reflects the fact that phenylpyruvate, which belongs to a general class of chemical substances known as phenylketones, is excreted in the urine of affected people. This provides a simple method of diagnosis, and the disease is treated by carefully controlling the diet so that it provides no more phenylalanine than is needed for normal health. There is then no surplus to be converted into phenylpyruvate.

Phenylketonuria, therefore, can be treated with a high degree of success, but if left untreated it provides a clear illustration of Mendelian inheritance. One enzyme (phenylalanine 4-monooxygenase) is involved in the conversion of phenylalanine to tyrosine, and is coded for by one gene. A normal person has two good copies of this gene, one derived from the mother and the other from the father. However, a small proportion of the population, *heterozygotes* for the phenylketonuria gene, have only one good copy, the other one coding for something incapable of catalyzing the reaction. These people are also normal and healthy, and if one of them marries a person with two good copies, a *homozygote*, all of their children will also be normal and healthy. However, if two heterozygotes marry there is a one-quarter chance that both will pass on the defective copy to any one child; thus approximately one-quarter of the children of heterozygotes have phenylketonuria (Figure 9.1). All in all, phenylketonuria behaves just like one of the seven traits that Mendel studied in his pea-crossing experiments.

Nowadays we regard Mendelian genetics and evolution by natural selection as inextricably bound up with one another, but it was not always so. Although Darwin and Mendel were contemporaries (and Mendel was aware of Darwin's

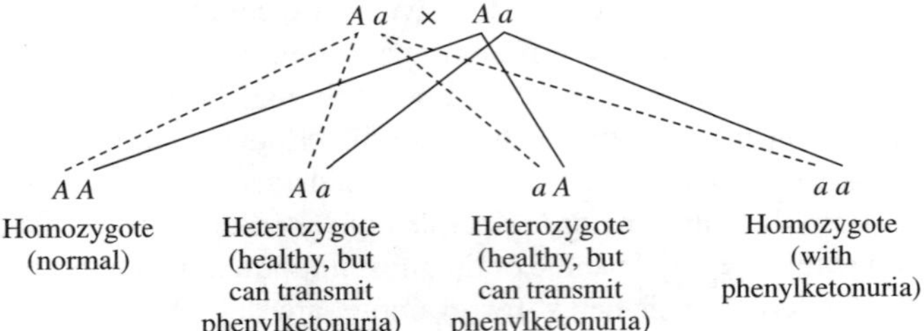

Fig. 9.1 Inheritance in diploid organisms. For the disease phenylketonuria, a normal homozygote has two good copies *AA* of the relevant gene, and a homozygote with the disease has two bad copies *aa*. A heterozygote *Aa* is healthy but can pass on the disease to children. The child of two heterozygotes has one chance in four of being a healthy homozygote *AA*, two chances in four of being a heterozygote, and one chance in four of being an abnormal heterozygote *aa* with phenylketonuria

work), Darwin, in common with the rest of the scientific world of his time, worked in ignorance of Mendel's experiments, and he thought of inheritance in paint-mixing terms. This misconception almost resulted in complete abandonment of natural selection as a credible theory, because in terms of paint mixing it was impossible to answer a criticism raised by an engineer named Fleeming Jenkin.

Jenkin's argument can be understood by examining how pea breeding actually proceeds (under Mendelian rules) and how it would proceed if it followed paint-mixing rules. Suppose that I am a pea grower with a large number of plants with yellow seeds and a single plant with green seeds, but I am anxious to have just green seeds.

In the first generation I must have no option but to cross the green-seed plant with as many yellow-seed plants as I can, and make up the population by crossing yellow-seed plants with one another. On the true (Mendelian) model this will give me some heterozygotes, all of them with seeds that look green to the eye (remember that green is dominant over yellow), and some homozygotes with yellow seeds. If the population size is large, the yellow-seed homozygotes will constitute the large majority (unless I am willing to risk the possibly catastrophic fall in population size that would result from avoiding all yellow–yellow crosses, even in the first generation). Nonetheless, in the next generation I can cross all the plants with green seeds with one another, and will need to use fewer plants with yellow seeds. One-quarter of the green–green crosses will yield plants with yellow seeds in the second generation of crosses, which I can discard, and three-quarters will give green seeds. Of these, two-thirds will be heterozygotes and one-third will be homozygotes, but if I am working in the nineteenth century I have no way of knowing which are which and will have to use green heterozygotes and green homozygotes indifferently.

No matter: I have many more plants with green seeds, including many more homozygotes, than I started with. Suppose that in each generation I can obtain four offspring from any pair of parents (pessimistic, but it allows simple calculations). Crossing the original green-seed plant with a yellow-seed plant will then give me four heterozygotes in the first generation, which will give me, on average, four yellow-seed homozygotes in the next generation, eight heterozygotes and four green-seed homozygotes. In two generations, I have increased the number of plants with green seeds 12-fold, and the number of heterozygotes 4-fold. It becomes a little complicated to calculate exact numbers after this, because we need to allow for all the possible crosses we can make in each generation. Nonetheless, it is clear that after the first generation I am always crossing green-seed plants with green-seed plants, so the proportion of plants with green seeds increases by a factor of between three and four each time, and as I discard all plants with yellow seeds other than those needed initially to maintain the population, the proportion of homozygotes among the plants with green seeds also increases in each generation. So, even if the detailed calculation is complicated, it is not difficult to see that in a few generations I can produce a population consisting overwhelmingly of plants with green seeds. I may continue to produce a small proportion of plants with yellow seeds for many generations, as in the absence of modern genetic techniques there is no way to recognize the heterozygotes until it is too late, but this is a minor nuisance.

Suppose the problem were the opposite: that I started with a population consisting overwhelmingly of plants with green seeds, with just one plant with yellow seeds, but I wanted to have just yellow seeds. In one sense this may appear more difficult, as the first crossing of yellow-seed plants with green-seed plants will give me no plants with yellow seeds at all. After that it becomes very easy, however, as now it is easy to recognize the heterozygotes: any green-seed plant with a yellow-seed parent must be a heterozygote. So in the second generation I produce as many offspring as I can by crossing heterozygotes, and one-quarter of their offspring will have yellow seeds. Once I have even a few plants with yellow seeds I can cross them exclusively with one another and the proportion of yellow-seed plants then increases rapidly in each generation.

Selecting for a characteristic that I want is thus relatively easy in the real, Mendelian world, regardless of whether the gene that interests me is dominant or recessive. Matters would be very different in a paint-mixing world. In such a world it would not matter whether I wanted green seeds or yellow, so we can return to the original form of the problem where all plants but one had yellow seeds, and I preferred green. What could I do? Clearly I would start as before, by crossing the lone green-seed plant with a yellow-seed plant, but now I would get not four plants with green seeds but four with greenish-yellow seeds. If I crossed these with one another I could obtain other plants with the same characteristics, or if I crossed them with yellow-seed plants I could get plants with seeds more yellow than green, but there is no way I could get a plant with green seeds. With completely unnatural selection, the very best I could hope for in the long term would be a population with greenish-yellow seeds. With natural selection without rigorous elimination of the undesired characteristics in every generation, the results would be much worse: a few plants with greenish-yellow seeds in the first generation, which would rapidly dilute into plants indistinguishable from the original population after a few generations.

Darwin spent a great deal of time in his later life meeting and talking with pigeon breeders, and he was well aware that they could actually produce an amazing variety of pigeons, in contrast with the gloomy view of their prospects that the paint-mixing model suggests. Dogs provide perhaps a better example for people more familiar with them than with pigeons: in the wild they are as uniform as any other wild species, but unnatural selection can produce varieties as different as the pekinese, the dalmation, the St Bernard, or the chihuahua. As R. A. Fisher pointed out, an even more striking example stares us in the face every time we see a human family: the child of a man and a woman is typically not of intermediate sex but is either completely boy or completely girl. Clearly sex is not a character that can be mixed like paint, but is inherited as a whole.

However, we do not always see the things that are staring us in the face, and although by Darwin's time breeders of pigeons, dogs, agricultural animals, and plants had acquired a great deal of practical knowledge of how to obtain the results they wanted, any theory that guided them was completely false. Until Mendel pointed it out, no one appeared to have noticed that inheritance was particulate. Even then, no one took much notice of Mendel, and when eventually they did, at the beginning of the twentieth century, they mostly took his work not as explaining how natural selection could work, but as making it unnecessary as a mechanism for evolution. As late as 1932, T. H. Morgan, in his influential book *The Scientific Basis of Evolution*, regarded natural selection merely as an effective purifying mechanism for eliminating harmful mutations. This attitude was standard for its time, but it seems so perverse today that it is hard to understand how it came about.

It is also hard to understand why so little attention was paid to Mendel's work during his lifetime, when he had provided the answer to questions of widespread interest and importance. The neglect is often attributed—sometimes explicitly, but more often by implication—to his supposed isolation from the world in the Augustinian monastery at Brno in Moravia. However, this image of an unworldly monk tending his pea plants is completely misleading. Mendel was in reality one of the first and most successful examples of someone bringing a training in mathematics and physics to bear on a biological problem. He had had an excellent university training in these subjects, first at Olomuc and later at Vienna, where his teachers included the distinguished physicist Christian Döppler.[1] His results were by no means unknown during his lifetime: he corresponded with the great scientists of his day, and his published work

[1] Döppler is remembered today for explaining the change in tone that you hear when an express train passes close by. He arranged for a brass band to play while seated on a moving train: listeners in the station heard the music as sharp while the train was approaching, but as flat when it had passed. In English this is usually called the *Doppler effect*.

was widely circulated. The failure of his contemporaries to appreciate his discoveries must be interpreted in terms of lack of understanding, not lack of information.

Returning to metabolic systems, we saw in discussing them that although stability is a desirable feature we had no need to call on natural selection to explain it, because just assigning kinetic parameter values at random to all the enzymes in a system would result in a stable steady state much more often than not. We also saw that all enzymes would appear to be present "in excess" if we looked at them in too simple-minded a way. According to the summation theorem discussed in Chapter 8, the control of the flux through a pathway is shared by all of the enzymes in the pathway, and it follows that the average share held by any single enzyme is of the order of one divided by the number of enzymes: if there are 10 enzymes, the average share is one-tenth, and so on. It is slightly more complicated than this, because some of the "shares" can be negative, allowing some of the positive ones to be larger than a simple analysis would suggest, but, although this represents a real complication in some circumstances, it is usually unimportant enough to be ignored.

How many enzymes need to be considered depends on how large a system we try to analyze. At the grossest level we can regard the system as the entire organism, and the flux as just the rate of growth. For growth of the whole organism we must be talking about thousands of enzymes, so the average share of the rate of growth held by any one of them is less than one part in a thousand, and this means that we can alter the activity of a randomly chosen enzyme by quite large amounts without any detectable effect on growth at all. There is, however, a limit to this: if the particular enzyme activity is absolutely essential to the life of the organism and there is no alternative way of providing the essential function, then eliminating the enzyme entirely will certainly have an effect; the organism will be unable to live. In practice, as I shall discuss in the last chapter, this happens less often than one might guess, because organisms usually do have alternative ways of achieving any particular function. In organisms where the question has been studied, therefore, organisms as diverse as mice and yeast, only about one out of every five genes is absolutely essential in this sense.

Even if we are talking about a more specific character than the capacity to grow, for example the color of the seeds of a plant, then the pathway responsible for producing the character, typically a set of reactions leading to a dye, will usually contain several enzymes. So we should expect to see several enzymes that share most of the control of the rate at which the dye is produced. One of these enzymes may have a flux control coefficient of one-fifth for production of the dye, meaning that if its activity is decreased by 5% the dye will be produced about 1% (one-fifth of 5%) more slowly. This will certainly not be noticeable without accurate measurements. Extrapolating to larger changes is not very accurate (because, in general, flux control coefficients become larger as the enzyme activity concerned becomes smaller), but it remains fair to estimate that a decrease to one-half of the normal activity of the enzyme will

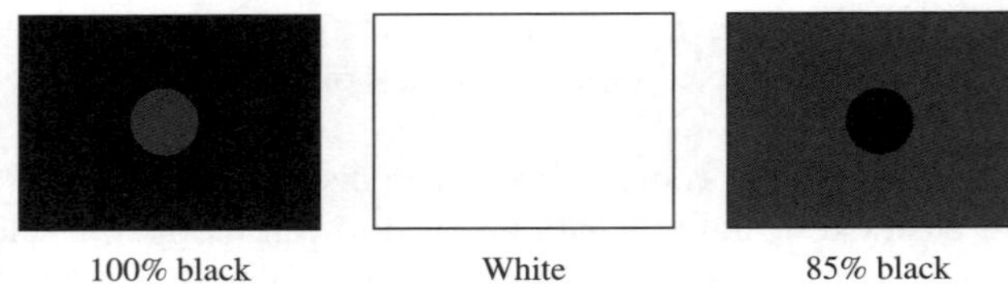

Fig. 9.2. Small differences in phenotype. The outer rectangles look virtually the same (and quite different from the white rectangle in the middle) even though the left-hand one is 100% black whereas the right-hand one is only 85% black. The difference is easier to see when the two shades touch one another: the central circle at the left is 85% black, and the one at the right is 100% black

decrease production of the dye by between 10% and 15%—big enough to be detected quite easily with instruments, but small enough to be missed in a judgment by eye (Figure 9.2). However, if we decrease its activity to zero, and if there is no alternative way of making the dye, the amount of dye produced must fall to zero as well.

These simple arguments mean that regardless of any genetic considerations there is a huge difference between decreasing an enzyme's activity by one-half and decreasing it to zero: in the one case the effect may easily pass unnoticed; in the other some metabolic product will not be made—in the one case peas that are near enough to an average green to fall within the expected scatter for any biological variable; in the other case peas with no green dye at all, which will be seen as yellow.

This is, in essence, the explanation that Henrik Kacser and Jim Burns offered in 1981 of why nearly all mutations in diploid organisms are recessive; it differs from Fisher's explanation in being easy to understand and so obviously right that any experimental test may seem superfluous. However, it is always a good idea to devise an experimental test, even of something that appears obviously right, especially if it goes against the established wisdom in the field accumulated over half a century. This was done by Allen Orr, who took advantage of the capacity of *Chlamydomonas reinhardtii* to reproduce occasionally as a diploid even though it spends the overwhelming proportion of its time as a haploid. Moreover, unlike some other organisms, such as fungi, its haploid state is truly haploid, with no more than one nucleus per cell, and no more than one copy of each chromosome. Thus it has no opportunity to experience multiple copies of a gene except during diploid generations. Even if Fisher's explanation of dominance is correct, therefore, *Chlamydomonas* has too few opportunities to benefit from its occasional diploid moments to show any effect of natural selection exerted on diploid cells. On Fisher's hypothesis, mutant genes in *Chlamydomonas* should be recessive much less frequently than they are in species that are always diploid and can experience natural selection of Fisher's modifier genes in every generation. No such effect is observed, however: Orr found that the great majority of mutations in *Chlamydomonas* are recessive when studied in diploid cells, exactly as found by Fisher in his original study of the fruitfly *Drosophila*, a normal diploid species.

This result leaves no escape for Fisher. If a pattern that requires selection in diploid (or polyploid) cells to work proves to be exactly the same in a species that is nearly always haploid, the explanation cannot be right. As noted by Orr, his observations also dispose of two theories proposed by J. B. S. Haldane, one of which is similar to Fisher's though (for me at least) much easier to understand. This theory supposes that natural selection has a tendency to compensate for the harmful effects of mutations by replacing wild-type alleles that produce "just enough" enzyme with ones that produce "too much." Now, as we saw in Chapter 8, the whole idea of having "too much" of most enzymes is a misconception (though an excusable one for Haldane, who was writing more than 70 years ago): just about any mixture of enzymes catalyzing a series of linked reactions will result in a state in which there is apparently more than enough of each enzyme. However, as Orr noted, there is a different reason for rejecting this explanation of dominance. Like Fisher's it requires selection in heterozygotes, and therefore cannot explain why mutant genes are just as likely to be recessive in the very rare heterozygotes of a principally haploid species like *Chlamydomonas* as they are in species that are always diploid.

This result parallels the theme of the whole book, and illustrates that even as great a geneticist as Fisher can be guilty of finding an adaptation in a phenomenon that requires no evolutionary explanation, because it follows necessarily from the diploid pattern of inheritance.

10
An Economy that Works

Xenophon is mostly remembered today as the author of the *Anabasis*, where he records the story of the 10 000 Greek soldiers from Cyrus's army that he led back home after the unsuccessful war with Artaxerxes. However, he has other calls on our attention, the most relevant to this book being that he wrote the *Oeconomicus*, probably the world's first textbook of economics. It contains ideas that formed the basis, more than 2000 years later, of the first serious attempt to analyze economic phenomena, which was made by Adam Smith in *An Inquiry into the Nature and Causes of the Wealth of Nations*. Xenophon noted that there is a relation between the number of blacksmiths in a region and the prices that they can charge: if there are too many, the price of their work falls, causing some to go out of business and helping to restore the value of the work of those that remain. This is effectively the first expression of what we now call the *law of supply and demand*.

However, no observer of modern economic systems can fail to be struck by the fact that economic laws do not seem to work very well. This is in part due to the tendency of governments to think that the laws do not apply to them and that they can manipulate them for their own purposes. We shall examine other possible explanations at the end of the chapter. Meanwhile we need to look at a domain in which the laws of supply and demand work supremely well, namely the organization of metabolism in a healthy organism. This is an aspect of metabolism that has interested me for a long time, and in the past 15 years or so I have been working with Jannie Hofmeyr, a biochemist in Stellenbosch, South Africa, to explore the relationships between economics and metabolism. Our aim has been to develop a modern theory of metabolic regulation that pays proper attention to ideas of supply and demand.

The metabolic factory needs to synthesize a vast array of different products to satisfy all of its activities, and each of these is made in just the right quantities at just the speed needed. (In crises, of course, such as may follow a physical injury or infection by a microbe, sudden unexpected demands for particular metabolites may exceed the capacity of the organism to satisfy them, and in extreme cases this may result in death, but I am speaking here of periods of health. In any case, organisms typically respond far more efficiently to crises than human societies do.)

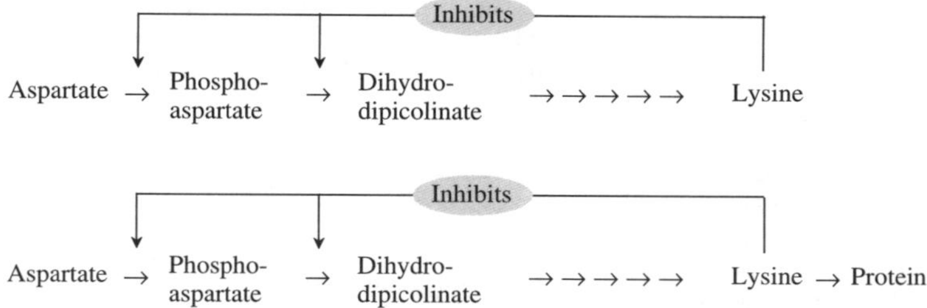

Fig. 10.1 Biosynthesis of lysine. In textbooks metabolic pathways are typically drawn with no indication that the end product (lysine in this example) is synthesized in order to be used (top). In fact, lysine is needed for protein synthesis (bottom), and if this essential step is not taken into account it is impossible to rationalize the regulation of the pathway

The organization of much of metabolism according to supply and demand is obscured in most biochemistry textbooks by omitting the demand component of a metabolic pathway as it is usually drawn; in consequence, this essential aspect of the organization passes unnoticed by most biochemistry students. Thus "everyone knows"—every biochemist, anyway—that a metabolite such as lysine is made in order to be used as a building block for protein synthesis. But a textbook illustration of lysine biosynthesis in bacteria such as *Escherichia coli* will typically stop at lysine (Figure 10.1). Worse still, it is typically called an "end product," or even sometimes an "ultimate end product." This may not matter very much if the aim is just to illustrate the chemical reactions that are used to transform another aminoacid, aspartate, into lysine. After all, no one expects an account of the processes used in a factory that makes shoes to include a description of how the shoes are taken off to be sold and worn by consumers—"everyone knows" that this is what happens to shoes after they have been made. Nonetheless, an account of the economics of shoe factories would be regarded as seriously deficient if it dealt only with the methods of production, entirely ignoring the demand for the finished product.

So it is with biosynthetic pathways: as long as we are talking just about the chemical transformations, it is of no great account if we ignore what happens afterwards. If we also ignore demand when we ask how the pathway is regulated, however, we make a serious error because our description then omits the key to the whole business. Unfortunately, most accounts of metabolic regulation have done just that since biochemical feedback was discovered in the 1950s, and although the principles are not difficult they are frequently presented so badly that they remain very poorly understood half a century later.

If a bacterial cell needs more lysine, say, for protein synthesis it simply uses more. This causes a transient fall in the concentration of lysine, which is sensed by enzymes at the beginning of the pathway that transforms aspartate into lysine, making them more active. Conversely, when the cell needs less lysine it just uses less, causing a transient increase in the concentration of

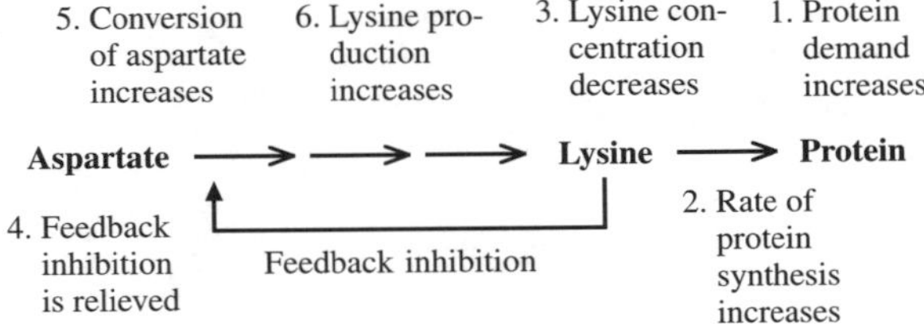

Fig. 10.2 Regulation of lysine biosynthesis by feedback inhibition. The explanatory notes 1–6 within the illustration are intended to be read in the order of the numbers

lysine, which has the opposite effect of making the same enzymes at the beginning of the pathway less active. The way the mechanism works is illustrated in Figure 10.2. It is called *feedback inhibition*, or sometimes *allosteric* inhibition, for a reason that we shall come to shortly.

This regulatory system works just like the negative feedback systems used by engineers to control the output of regulatory devices, such as the thermostat in a refrigerator. In the latter case, the explanation in an engineering textbook will make it quite clear that what is being regulated is not the flow of heat but the temperature. A thermostat does not prevent you from leaving the refrigerator door open, but it can *respond* to the increased heat flow if you do by working harder to keep the temperature constant. By contrast, although biochemistry textbooks will often *describe* the regulation of lysine synthesis in much the same way as I have just done, they will often confuse the interpretation by implying that it is the metabolic flux that is being regulated, whereas the regulatory design is a design for regulating concentrations, not fluxes.

Before continuing, we need to ask whether we are really entitled to attribute the existence of feedback regulation in metabolism to natural selection. Maybe it is just an inevitable consequence of the organization of metabolism into series of linked reactions, like some other properties mistakenly attributed to natural selection that we discussed in Chapters 8 and 9. To decide this we should compare metabolism with other flow systems found in nature or engineering. In natural river systems, for example, there is no law of supply and demand. The existence of a drought around the lower reaches of a river does not cause the tributaries at the higher levels to flow any faster; the existence of a flood does not make the snow melt more slowly in the mountains. On the contrary, a river is a water delivery system that is completely indifferent to the "needs" of the plants and animals that live at the lower levels.

What about artificial water supply systems? Maybe just enclosing the flow from reservoir to final user in watertight pipes is sufficient to ensure that the laws of supply and demand will apply. We can imagine a pipe leading from a reservoir situated in the hills 1000 m above the user, open at the level of the reservoir but terminated by a tap at the level of the user. Opening the tap will make the water flow; closing it will make it stop. Not only that, but we can obtain a less-than-maximal flow by opening the tap only partially.

Surely we have here a satisfactory flow system that obeys the laws of supply and demand and requires only the crudest of "designs." Not so. Although such a system might appear to work as a design on paper (if we refrain from asking any questions about the pressure inside the pipe), it would work very badly in practice, if at all, as any civil engineer would realize immediately. A head of water 1000 m in height implies a pressure of about one hundred times atmospheric pressure at the level of the tap, when it is closed and there is no flow of water. To resist this, the pipes would need to be extremely strong. Moreover, as we can hardly conceive of a single unjointed pipe leading all the way from the reservoir in the mountains to the tap at the other end, there would need to be separate portions of pipe with joints between them, and at the point of highest pressure there would need to be a joint between the end of the pipe and the tap. All these joints would need to be extremely strong and completely watertight, and any repairs that would imply opening the system would be impossible to undertake because of the powerful jets of water that would spurt out at any point of opening.

Worse still, the pressure at each joint would not be a constant static pressure, but it would fall drastically each time the tap was opened and the water started flowing, only to increase again when the tap was closed. So the joints would need to withstand the changes in pressure. The tap itself would be very difficult to manipulate because, when closed, it would be under one hundred times atmospheric pressure: this is the pressure that bears down on an area of 1 cm^2 carrying a weight of 100 kg (or an area of 1 $in.^2$ carrying a weight of more than half a ton); if you prefer, it is the pressure that a submarine feels at a depth of 1 km.

So, although a sealed tube connecting a reservoir to a tap might work quite effectively on paper as a way of regulating the flow of water, it would fail in practice because the only way in which the flow would be regulated at the point of exit from the reservoir would be by the back pressure in the tube, and, as we have seen, this would vary so much, and go to such large values, that it would result in intolerable stresses on the whole system. In practice, civil engineers solve this problem by ensuring that the final point of demand is not the only place where the flow is regulated. At the very least, there needs to be a feedback loop from the point of demand to the point where the supply is initiated. In other words, the exit from the reservoir needs to receive information about the demand that comes by a route that avoids the need for the information to be transmitted solely by the pressures in the pipe. If there are several different points of demand (as there normally will be in any water-supply system), there needs to be a separate regulatory device at each exit from a point where the supply pipe divides into two or more smaller ones: you do not want to be deprived of the possibility of taking a shower because your neighbor is watering his garden. This can of course happen in real water (or gas, or electricity, or whatever) supply systems, but it is a sign of a design that is insufficient to meet the demand. In a well-designed and well-maintained system, your water supply does not depend on how much water your neighbor is using.

Notice that although the simple sealed tube with a tap at the end requires (in principle) no special design, and has properties that follow automatically from its structure, the same is not true of systems with feedback loops. There is nothing automatic or inevitable about a signal that passes around all the plumbing to inform the regulators at the reservoir about changes in demand: this must be a design feature deliberately installed by the engineer.

Nearly all of this analysis applies equally well to metabolic regulation. The only important difference is that the water system that supplies your home was planned and designed by real engineers with conscious intentions, whereas the mechanisms that regulate metabolic systems were not planned at all and their "design" is just the result of natural selection of the designs that work well, with rejection of the ones that do not. However, the explanation here is just the same as the one that Darwin and his successors apply to all suggestions of design in biology, and there is no need to labor the arguments here. Suffice it to say that metabolic regulation, like any other biological adaptation, has arrived at a state that looks as if it had been consciously planned by an engineer.

The basic features were discovered in the 1950s, and many examples were examined in the next two decades that support the generalization that each metabolic pathway is regulated by "feedback inhibition at the first committed step." What does this mean? *Feedback inhibition* means that an enzyme acting early in a pathway is inhibited, or made less active, by accumulation of a metabolite that occurs late in the pathway, and the *first committed step* is the first reaction after a branchpoint that involves a large change in energy. Essentially this is just like what we discussed in the water supply: signals that measure the demand for end-product loop around the series of chemical reactions to act directly on the supply at an early enough point to avoid large variations in metabolite concentrations between the two.

We cannot carry out an experimental test of what would happen if feedback inhibition did not exist at all in a living organism, because the organism would not be able to support the stresses that this would produce, even if we knew how to make all the genetic modifications necessary to create such an organism. The best we can say is that if gene manipulation is used to suppress the feedback inhibition in just one pathway in a bacterium this can create difficulties for the cell: poor growth, leakage of metabolites through the membrane, etc., which can be interpreted as resulting from inadequately controlled variations in concentrations. Sometimes, it must be admitted, the effects are less dramatic than one might guess, but this probably just reflects the incompleteness of our knowledge and understanding. For example, in a yeast strain in which the enzyme phosphofructokinase lacks the regulatory controls believed to be important in the normal strain, metabolic fluxes and growth are normal; however, some concentrations of intermediates are abnormally high, and the strain responds to changes in conditions more sluggishly than the normal strain.

On the other hand, in the computer it is easy to model what would happen in an organism with no feedback controls at all, and it turns out that a pathway without feedback can vary its flux quite satisfactorily in response to the demand, but it does so at the expense of huge variations in metabolite concentrations. In other words, it behaves just like the water supply in a sealed tube without feedback.

For a water supply we can, if we are interested, examine the history of its construction and confirm that the regulatory controls that it contains were put there deliberately and are not just haphazard characteristics of its structure. Can we be equally sure that the corresponding features in metabolic regulation are the result of natural selection and do not follow automatically from the properties of all enzymes? After all, enzyme inhibition is a common enough phenomenon that can be exerted by many small molecules without the need to postulate a biological function. One answer might be that although feedback inhibition is common, the opposite phenomenon of feedback activation is rare almost to the point of nonexistence. However, this would not be altogether convincing, because all forms of activation (including feedforward activation, which could fulfill a plausible physiological function in some circumstances) are less common than inhibition. On the other hand, feedforward inhibition, when an enzyme late in a pathway is inhibited by a metabolite that occurs early in the pathway, is also extremely rare, and there would be no reason to expect this if feedback inhibition were a haphazard property.

Moreover, if we can claim that feedback inhibition occurs in metabolism because it fulfills a useful stabilizing role, we can also claim that one of the reasons why feedback activation is very rare is that if it occurred it would be dangerously destabilizing. Feedback activation occurs in everyday life when a microphone is placed within the range of a loudspeaker that it feeds, and the howling noise that we hear in these circumstances is a symptom of a system out of control.

After mention of feedback inhibition and activation, as well as feedforward inhibition, it will be evident that there is a fourth possibility—feedforward activation. This is also rare, but, significantly, when it does occur it is sometimes in pathways that are *not* demand-driven. We shall meet an example of this shortly when we consider glycogen production in the liver.

In addition, feedback inhibition has two other characteristics that would be unlikely to arise haphazardly: in longer pathways there is often very little structural resemblance between the end product that inhibits and the substrate of the inhibited enzyme; and the inhibition is usually "cooperative." Let us consider what these two properties mean. When inhibition arises for no obvious biological reason, it is usually for an obvious enough chemical reason: the substrate and inhibitor are similar enough in terms of chemical structure that the inhibitor can bind to the same site on the enzyme as the one where the substrate binds; however, as it lacks some feature necessary for the chemical reaction it does not react but does nothing. A classical example is provided by the enzyme succinate dehydrogenase, which uses succinate as its substrate but is inhibited by malonate (Figure 10.3). Succinate and malonate have almost the same chemical structures, so either is likely to bind to a site intended for

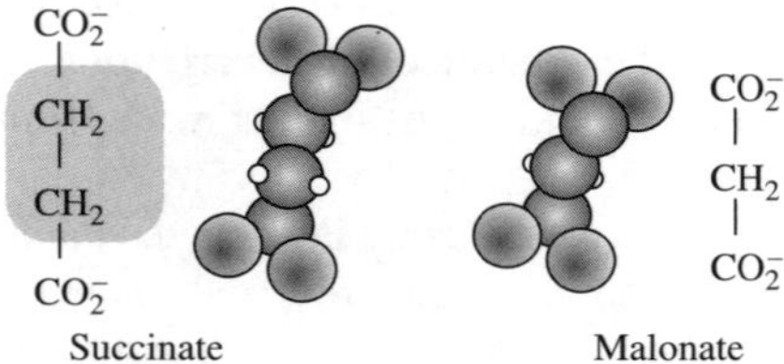

Fig. 10.3 Substrate and inhibitor. The reaction catalyzed by the enzyme succinate dehydrogenase alters the $-CH_2-CH_2-$ grouping (shown with a gray background) in succinate. Malonate has a similar size and structure, but does not have this grouping: it interacts well with the groups on the enzyme that allow succinate to bind, and so it can be bound to the site of reaction; but it cannot undergo the reaction, so it is an inhibitor rather than a substrate

the other, but malonate lacks the particular carbon–carbon bond that is transformed in the reaction catalyzed by the enzyme, so it does not react.

By contrast, feedback inhibitors may be quite different in structure from the substrates of the inhibited enzymes. For example, the aminoacid histidine is made in some bacteria from a sugar derivative called phosphoribosyl pyrophosphate. This is a far from obvious transformation, several chemical steps being needed to connect them, as histidine and phosphoribosyl pyrophosphate are structurally very different, and there is no chemical or structural reason for histidine to bind to a site intended for phosphoribosyl pyrophosphate. Yet the enzyme that transforms phosphoribosyl pyrophosphate is indeed inhibited by histidine. Moreover, the two do not bind to the same site: the inhibitory effect can be destroyed by poisoning the enzyme with mercury, but this poisoning does not affect the catalytic activity, and nor does it affect the capacity of histidine to bind; it just destroys the connection between the site where histidine binds and the site where the reaction takes place. This type of inhibition by a molecule that does not resemble the substrate is called *allosteric inhibition* (from a rather rough translation into Greek of the idea of a different shape), and is characteristic of metabolic regulation. Allosteric inhibition is quite common in circumstances where it has an easily identifiable biological function, and virtually unknown otherwise. Thus it can only be a design feature produced by natural selection, and not a random property of the enzymes affected.

The other property that can only be reasonably interpreted as a design feature is cooperativity. This describes the capacity of some enzymes to respond more sensitively to biologically significant inhibition than the general run of enzymes do to inhibition in general. For properties that arise just from the inherent structure of enzymes, the dependence of the rate on the concentration of substrate or inhibitor that modulates it is rather sluggish. Typically, to increase the rate of a reaction from 10% to 90% of the maximum by varying the substrate concentration, the latter needs to increase 81-fold (Figure 10.4). For inhibitors the relationship is a bit more complicated, but essentially the

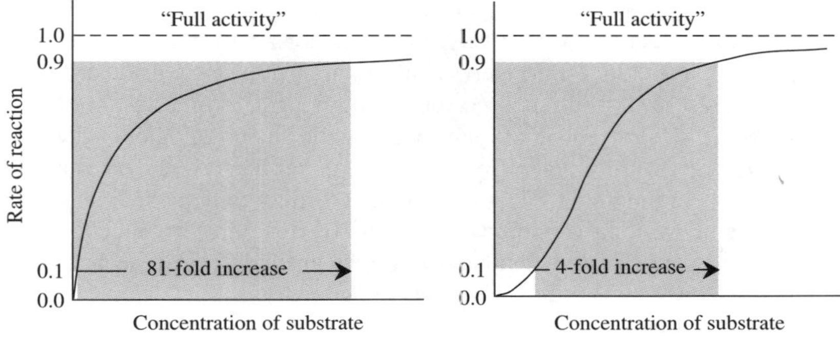

Fig. 10.4 Cooperative binding of a substrate to an enzyme. In a simple enzyme with no special regulatory properties (left) the activity depends on the substrate concentration according to a curve that requires an 81-fold increase in concentration in order to pass from 10% to 90% of full activity. Some enzymes use a property known as cooperativity (right) to convert this into an S-shaped curve, so this ratio is substantially decreased, to three or fourfold

same idea applies: to decrease the rate from 90% to 10% of what it would be in the absence of the inhibitor, the inhibitor concentration needs to be increased by a large factor, not necessarily exactly 81-fold but of that order of magnitude.

This is clearly not very satisfactory as the response to a signal. Imagine that traffic signals worked not by changing the color from red to green but varying the intensity of a single light, so that high intensity meant stop and black meant go. Suppose further that cars were made so that to go from 90% to 10% of the maximum velocity, the light from the signal had to increase 81-fold in intensity. Clearly it would not work. You might be willing to put up with a car that could never go faster than 90% of its theoretical maximum (most of us do drive cars that cannot go nearly as fast as the maximum indicated on the speedometer), but a car that could not go at less than 10% of the maximum would be intolerable.

Likewise in metabolism, enzymes that need to respond sensitively to signals do so *cooperatively:* this just means that they are more sensitive than they would be if they just had the standard off-the-shelf properties that any enzyme has. The two oxygen-binding proteins that we have met already in Chapters 2 and 3—hemoglobin and myoglobin—illustrate the difference very nicely. They are not enzymes, as they do not catalyze a reaction, but hemoglobin was so useful during the development of our understanding of cooperativity that it was often called an "honorary enzyme": even though it does not catalyze a reaction, it does share the properties of cooperative enzymes that are relevant to the discussion. In particular, filling hemoglobin from 10% to 90% of capacity with oxygen requires much less than an 81-fold increase in the oxygen pressure. The actual ratio varies somewhat with the conditions of measurement, but is of the order of fivefold, very much less than 81-fold. As hemoglobin has to be able to take up oxygen in response to relatively small variations in oxygen pressure in the lungs, and release it in response to relatively small variations in oxygen pressure in the tissues where it is used, this sensitivity fulfills an obvious biological need.

By contrast, myoglobin is used for storing oxygen in muscles. To do this effectively, it needs to bind oxygen more tightly than hemoglobin does, so that it can strip off the oxygen from the hemoglobin that arrives, but it has no need of cooperativity as it releases oxygen according to demand over a very wide range of working conditions. Accordingly, therefore, we should not be surprised that oxygen does bind more tightly to myoglobin than it does to hemoglobin, and that it does so without cooperativity.

These two proteins also illustrate quite nicely what structural features allow proteins to achieve cooperativity. Hemoglobin has four oxygen-binding sites on each molecule, and the term cooperativity comes from the idea that these sites cooperate with one another: once one binding site has taken up an oxygen molecule, interactions with the other sites make it easier for them to do so as well. By contrast, myoglobin has just one oxygen-binding site on each molecule and has no possibility of a corresponding cooperation between sites. So we have one protein with cooperativity and multiple binding sites, and another with just one site and no cooperativity, and together they illustrate the normal combination of structural characteristics and binding properties. Enzymes with just one binding site almost never display cooperativity (and in the rare exceptions the cooperativity has to be explained in more complicated ways than simple cooperation between sites). Enzymes with multiple sites may or may not display cooperativity, as the fact that they have the possibility of interactions between sites does not mean that they must have such interactions.

Like allosteric inhibition, cooperativity is not scattered haphazardly among the known enzymes without regard to any function that it might have, but occurs almost exclusively in enzymes where a regulatory function seems obvious, for example in the enzymes that catalyze the first committed steps of different pathways. In fact, even though there appears to be no structural feature of proteins that makes it necessary that an enzyme with an allosteric response to a particular metabolite must also have cooperative kinetics, or vice versa, they do in practice occur so often in the same enzymes that the two terms have often been regarded as synonymous. It would be thus perverse to regard the regulatory properties of enzymes as anything but an adaptation produced by natural selection.

Although, as I have suggested, the most common need that is satisfied by enzyme regulation is to ensure that the laws of supply and demand are obeyed as painlessly as possible—with as little variation in metabolite concentrations as possible—it would be wrong to suggest that this is the only need or that all pathways are necessarily designed in the same sort of way. Not all pathways exist to produce useful metabolites for the cell; some exist to get rid of harmful substances that come from toxins in the food or released by infective agents. Expecting these pathways—collectively known as detoxification pathways—to obey the same laws as biosynthetic pathways would be as absurd as requiring the sewage works in a city to operate under exactly the same rules as those that govern the supply of fresh water.

As far as I know no one has tried to develop a general regulatory theory for detoxification, but I should be surprised if ideas of supply and demand proved to be as prominent in it as they are in biosynthetic pathways. On the contrary, we should expect the major considerations to be getting rid of the harmful substance as fast as possible while making sure that anything it was converted into was either less harmful or at least could be removed rapidly. Given that detoxification pathways must deal with toxins that may not have been taken account of during evolution, we must expect them to make mistakes sometimes, especially with chemical poisons that may never have occurred in the natural world, applying a routine transformation that turns out to convert something mildly unpleasant into something much worse. Although, in general, it may be a good bet to start working on a water-insoluble hydrocarbon from the petrol industry by oxidizing it to something that can be dissolved in water, in specific cases this may not be a good idea at all, and some of the more potent cancer-producing agents are produced by the body itself in its efforts to detoxify foreign substances.

We have seen various examples already of harmful substances produced by metabolism itself. In phenylketonuria, described in the previous chapter, the major harm is done by a molecule produced by the body in its efforts to remove the excess of an aminoacid that is not harmful at all when present in normal amounts. We also encountered the enzyme catalase (in Chapter 1), which exists to destroy the very dangerous chemical hydrogen peroxide that arises as an unavoidable side effect of using oxygen in the first stages of dealing with a wide range of unwanted molecules; superoxide dismutase, which we shall meet in the next chapter, has a similar role, in destroying an even nastier substance that arises as a side effect of reactions that use oxygen.

Even with a toxic substance like ethanol (ordinary alcohol), which is certainly not a new product from the chemical industry but has been present on an evolutionary timescale, it is chemically convenient to begin the detoxification by converting it into something worse, namely acetaldehyde. In such a case, of course, the regulatory design has to ensure that the acetaldehyde is removed as fast as it is produced. In this instance, medical practice has taken advantage of the toxicity of acetaldehyde by using antabuse, or disulfiram, as a treatment for alcoholism. This substance has very little toxicity of its own, in fact very little effect of any kind when there is no acetaldehyde in the system. However, it blocks the enzyme that catalyzes the removal of acetaldehyde. If there is no ethanol in the system this does nothing, but in the presence of ethanol the relatively mild symptoms of ethanol poisoning are replaced by the much more unpleasant symptoms of acetaldehyde poisoning—a throbbing headache, difficulty in breathing, nausea, copious vomiting, and sweating are the least of the problems. Patients administered antabuse are normally strongly advised not to touch alcohol, or even to use products like after-shave lotion that have ethanol in them, as long as the effect of the antabuse lasts.

We must also expect considerations of supply and demand to need some modification in the early stages of food processing. Herbivorous animals like cows and sheep eat almost continuously, partly from necessity, as the food they eat has so little nutritive value that they have no other way of getting enough, but also because they have no other more pressing things to do. Almost continuous feeding is also convenient for very young animals like neonatal rats even if their diet is richer in proteins than that of sheep. However, having to eat nearly all the time would be a nuisance for adults of many species, including our own (though a visit to a modern amusement park may suggest otherwise), and impossible for large carnivores like lions that tend to have large quantities of food available only at infrequent moments. In these cases, there is no advantage in being able to digest food very fast. On the contrary, it satisfies the requirements better if digestion of one meal takes a large fraction of the time available until the next meal. As a result, enzymes like pepsin, responsible for the first stage of digesting proteins in the human stomach, do not have any sophisticated feedback regulation but just operate most of the time at saturation. This is like having an open but narrow outflow from a large tank of water: the rate at which water flows out is more or less unaffected by how full the tank is (unless it becomes completely empty). The effect of this saturation of the protein-digesting enzymes is that the bloodstream of a meat-eating animal continues to receive a steady supply of aminoacids—the products of protein digestion—long after the last ingestion of protein.

Although carbohydrate eaten as starch is digested rather slowly, like protein, animals can also obtain quite large amounts in the form of small molecules like sucrose, glucose, and fructose. Sucrose is ordinary sugar as bought from a supermarket, and also occurs as such in some sweet plants. Glucose and fructose are natural components of sweet food like fruits and honey. Ingestion of any of these small-molecule sugars produces a very fast and potentially disastrous increase in the amount of glucose circulating in the blood. However, even if natural selection has not yet had enough time to respond to the existence of supermarkets and cheap sugar, it has had plenty of time to adjust to the availability of fruit and honey, and in a healthy animal disaster is avoided by a very rapid conversion of the excess glucose into glycogen, which was the subject of Chapter 6. This is stored temporarily in the liver, being subsequently released and converted back into glucose when the need arises.

This latter conversion back to glucose must certainly be demand-driven, like the biosynthetic processes I discussed earlier, but it is equally certain that the initial conversion of glucose into glycogen cannot be. It cannot be because the liver has no "demand" for glycogen, and, although it uses a little glucose for its own purposes, the amount is very small compared with the amount it processes, On the contrary, conversion of glucose into glycogen in the liver has to be supply-driven, so we may expect it to be an exception to some of the generalizations made above, and so it proves. Hexokinase D, the enzyme that catalyzes the first chemical step, is exceptional in a number of ways. Unlike other kinds of hexokinase that serve the physiological needs of organs like brain and muscle, it is unresponsive to the concentration of glucose 6-phosphate, the product of its reaction; in other words, it ignores the demand for product. In contrast, it responds cooperatively to the concentration of its substrate, glucose. Not only that, but even though glucose 6-phosphate does not inhibit hexokinase D, it does activate the later steps of conversion of glucose to glycogen in the liver. It is not by chance that this unusual example of feedforward activation occurs as a mechanism in an unusual example of a supply-driven pathway. Perhaps most striking of all, hexokinase D is as close as we can find to a genuine rate-limiting enzyme, something that I referred to as being largely a myth when I discussed it in Chapter 8.

However, enough of exceptions: there is little doubt that most of the processes that occur in metabolism need to satisfy the laws of supply and demand, which means that they need to respond to changes in demand and ignore small changes in supply. Moreover, there is abundant evidence that they are indeed designed in that way, but this has not prevented a large amount of wishful thinking. Failure to take account of the real regulatory design of metabolism is a major reason for the failure of the biotechnology industry over the past 20 years. Since genetic manipulation became possible at the end of the 1970s, a vast amount of money has been wasted—or "invested", if you prefer a more conventional word—on the search for the mythical rate-limiting step in each pathway that would allow commercially valuable metabolic products to be overproduced at will in suitable microorganisms.

Ethanol (ordinary alcohol), for example, has many industrial uses in addition to its presence in beverages and is produced on a huge scale. About half a million tons of glutamate are produced each year, and other aminoacids like lysine and methionine owe their tremendous industrial importance to their use by farmers to supplement corn as a feed for cattle, which is very cheap but does not contain enough of these essential components. Even carbon dioxide, which may seem to be just waste if we think of yeast fermentation only as a way of making wine (and mainly as waste in making beer), becomes the whole point of yeast fermentation in bread-making. These and other substances are produced on such a vast scale in the world's industry that even a small increase in productivity translates into many millions of dollars per year. The idea is that once you identify the enzyme that catalyzes the rate-limiting step you can overexpress the enzyme so that its activity in the living organism is higher than it would normally be; then the organism will make more product and everyone will be happy.

The major problem with this approach is that it does not work. In Chapter 9 we have seen some good theoretical reasons for believing that it cannot work, but let us also look at some experimental evidence. The crucial experiments were done by Jürgen Heinisch, Fred Zimmermann and their colleagues in Germany in the 1980s—early enough in the story for vast amounts of money to have been saved if the results had been taken seriously. One of the things that "everyone knows"—everyone who has followed a standard course of biochemistry, in this case—is that phosphofructokinase controls glycolysis.

We have met glycolysis already in Chapter 4 as a central metabolic pathway in virtually all organisms, essential for converting sugars into energy, and phosphofructokinase is an enzyme that catalyzes a reaction of glycolysis. Its activity is responsive to an impressive variety of metabolic signals that indicate the immediate requirements of the organism. This responsiveness is certainly of the greatest importance for the regulation of glycolysis, but it is translated in an excessively simple-minded way into the naive idea that phosphofructokinase is the rate-limiting enzyme and that increasing its activity will inevitably result in a higher glycolytic rate. However, when Heinisch increased the activity of phosphofructokinase by a factor of 3.5 in fermenting yeast, there was no detectable effect on the amount of alcohol produced.

Before we try to understand this, I must emphasize that this is an *experimental* result; it is not just the wild musings of an armchair biochemist. By now it has been repeated, not only in yeast but in other quite different organisms, such as potatoes, and over expressing phosphofructokinase in potatoes has just the same effect as it has in yeast: it has no effect. Nor does it have any effect in other organisms where similar experiments have been done.

To understand these results, we have to recognize that the whole point of metabolic regulation of biosynthetic pathways is to allow metabolites to be made at rates that satisfy the demand for them, without huge variations in the concentrations of intermediates. This last qualification is so important that I will say it again: adjusting rates is quite easy, and requires little sophistication in the design. The difficult part is keeping metabolite concentrations virtually constant when the rates change, and this is needed if the effects are to be prevented from leaking into areas of metabolism where they are not wanted.

When we refer to demand we are, of course, talking about the organism's own demand, not demands made by an external agent such as a biotechnologist. This seems such an obvious point that it ought hardly to be worth making, but as it is a point that often seems to have escaped the biotechnology industry altogether it is perhaps not as obvious as you might think. An organism such as *Saccharomyces cerevisiae*, the yeast used in baking, brewing, and wine-making, has been perfecting its metabolic organization for a long time—maybe not all the time since the appearance of flowering plants around 130 million years ago, because the yeasts used industrially are domesticated strains not found in the wild, but a long time nonetheless. Brewing is known to have been used in Egypt at least 4000 years ago, and even if that is much less than 130 million years ago, it still adds up to more *S. cerevisiae* generations than there have been human generations. It is the height of arrogance to think that a little tinkering with the expression levels of its genes will allow us to force as much glucose through its fermentative system as we would like. Its regulatory mechanisms have evolved for the benefit of *S. cerevisiae*, no one else. Of course, the yeasts used in the baking and fermentation industries have undergone a considerable amount of unnatural selection over the centuries, but this has always, at least until the past few years, involved selecting natural variants capable of growth under the conditions preferred by the technologist, not tinkering with the basic structure of the organism in the hope of obtaining quite unnatural properties.

As a general rule, therefore, if you try to overcome by brute force the regulatory controls built into any organism you will fail. On the other hand, with a little understanding of what these controls are designed to achieve, you may hope to achieve your aims by subterfuge. Once it is understood that most biosynthetic pathways are designed to satisfy the demand for end product and to be as unaffected as possible by changes in supply and in the work capacity of the enzymes involved in the biosynthesis, it should become obvious that manipulating either the supply of starting material, or the activities of the enzymes, will have no effect other than to stimulate the regulatory controls to resist. It is as futile as trying to stop the import of cocaine by interfering with growers or traffickers in Colombia: as long as the demand is there, it will be satisfied.

By contrast, if you could fool an organism by manipulating the demand for a desirable end product, you might persuade the regulatory mechanisms to help rather than hinder. An interesting example comes from the monosodium glutamate (MSG) industry in the USA. An important component of American cuisine (and also, incidentally, responsible for the "Chinese restaurant syndrome", a form of glutamate poisoning), this was produced by Merck during the 1950s by growing the bacterium *Corynebacterium glutamicum* (at that time called *Micrococcus glutamicus*) on pure cane sugar as a carbon source, a cheap source of sugar imported in large quantities from Cuba. The bacteria were deliberately deprived of biotin, a vitamin essential for making cell membranes, with the result that their membranes leaked glutamate, which could then be harvested by the food company.

At the end of the 1950s the company lost its cheap source of pure sugar for political reasons, and was forced to use molasses instead. Although this was sufficiently cheap it was much less pure and contained large amounts of biotin, which would have been far too expensive to remove. The bacteria no longer made low-quality membranes, which no longer leaked, so glutamate ceased being lost to the growth medium. To overcome the problem, a different way of making the membranes leaky was needed, and this was achieved by growing the bacteria in the presence of penicillin (which acts by interfering with the building of membranes).

The point of all this is that as long as the bacteria grew at all they produced enough glutamate for their own purposes, principally for making new proteins. If a large part of the glutamate made was being lost through their leaking membranes and ended up in soup for human consumption, they just made some more. That is what their regulatory mechanisms had evolved to achieve, and that is what they did. More generally, it follows that one of the most effective ways of causing a microorganism to produce more of a desirable metabolite will be to engineer a leak.

I do not know why the Merck engineers starved their bacteria of biotin in the first place. Maybe it was just because it was cheaper and they found that it worked. But whatever the reason, it is an example that is worth following. Engineering a leak, or finding some other way of artificially stimulating the demand for the desired product is likely to work as a general approach. Trying to force more metabolites through a pathway by increasing the supply or the activities of the pathway enzymes is just a way of triggering the regulatory controls that have been naturally selected to ensure that it will not work.

I started this chapter by noting that the classical laws of economics appear to work rather badly when applied to the domain for which they were devised, and indeed the title of the chapter is intended to reflect this idea. I continued by arguing that in the field of metabolism, in great contrast, they work extremely well. This may seem to present a problem: what difference between real economies and the economy of the living cell could justify believing that the same laws work badly in one and well in the other? I could reply that the living cell is the product of many millions of years of evolution and that even if not strictly a designed system, it shows all the characteristics of a designed system, whereas human economies are not designed and have not existed for long enough to have evolved to a state in which they appear to be designed. However, this is not a very satisfying answer, because some national economies, especially during the twentieth century, were very thoroughly planned, and to some degree therefore designed, but there is little to suggest that they worked any better than the rough-and-tumble system that prevails in the USA; most, indeed, would argue that they performed much worse.

I think the real answer lies in the fact that real economies are *complex* and metabolism is not. Enzymes and the other components of the cell are governed by physical laws, and human behavior, which determines how real economies behave, is not. In the cell, moreover, mass is conserved, but corresponding quantities in economies, such as money and the things it can buy, are not. However, we are still faced with a difficulty: what could be more complicated than a living cell, with its thousands of chemical reactions proceeding simultaneously, so how can I claim that it is not complex? In everyday conversation we often use the terms "complicated" and "complex" interchangeably, and ordinary dictionaries encourage this by giving each word as one of the definitions of the other, but in modern science they are not at all equivalent. A system is complicated if it contains many components, all of which need to be understood for the whole system to be understood. To be complex, however, a system needs more than that: it needs in some sense to be *more than the sum of its parts*, to display a property known as *emergence*, a manifestation of interactions between the different components that could never be deduced by examining them one at a time.

Now metabolic systems do not entirely lack such interactions. Some of the different enzymes do interact with one another and with components of the cell architecture that ensure that enzyme molecules are not evenly distributed around the compartments they find themselves in. Nonetheless, computer models of cell metabolism that ignore these complications manage to reflect fairly accurately, sometimes even quantitatively, the real behavior of the cell.

This means that even when complications exist, they act far more to stabilize than to destabilize the cell. The sort of positive feedback effects that make the behavior of real economies difficult to predict barely exist in metabolism. As we have seen in this chapter, negative feedback is very common in metabolism, positive feedback rare almost to the point of nonexistence.

Where complex properties occur in living systems, as for example in chaotic cardiac arrhythmias, they are characteristic of pathological failure of the normal controls—in this case in some kinds of heart disease—not of health. Even at the level of an individual enzyme, the enzyme peroxidase from horseradishes, an example of chaotic behavior has been known for many years, but we have no idea how its properties benefit the horseradish, and they contribute nothing to our present understanding of how metabolic systems are controlled in general. The sort of positive feedback loops that account for the complex behavior of economic systems, political relations, and indeed some aspects of biology such as the extravagant overdevelopment of the peacock's tail, are conspicuously absent from the major pathways in the metabolic economy. There is no reason, therefore, to see any contradiction in the claim that classical economic theory works much better in metabolism than it does in the domain in which it was developed.

Not everyone finds this analysis very congenial. Complexity is a fashionable subject at present, and the idea that metabolism is complex has considerable appeal, especially if you can see a Nobel prize beckoning. Given the real complicatedness of the subject, it becomes easy to slide into calling things complex when they are just complicated. Moreover, the noun "complicatedness" is such a mouthful that even people who are careful not to describe things as complex when they are just complicated will often use the same word "complexity" as the noun for both. However, one should resist: the subject is sufficiently difficult as it is without inventing difficulties that do not exist. In any case, even if much of metabolic regulation can be explained and analyzed without invoking complexity, some of it cannot, at least for the present, as there is genuine complexity in the organization of living organisms as well. Not all of the metabolic properties of a living organism can be understood in terms of the properties of its components, though not everyone agrees about the reasons. Some argue that it just reflects our current lack of adequate tools and knowledge, others that there is a fundamental barrier that can never be breached no matter how complete our knowledge becomes, and no matter how powerful our computers become. When the time comes to grapple with this in a serious way, it will not be helpful to find that we cannot use the natural word to describe it because it has been misappropriated to describe something else.

11
A Small Corner of the Universe

Consideration of the conditions prevailing in bisexual organisms shows that . . . the chance of an organism leaving at least one offspring of his own sex has a calculable value of about 5/8. Let the reader imagine this simple condition were true of his own species, and attempt to calculate the prior probability that a hundred generations of his ancestry in the direct male line should each have left at least one son. The odds against such a contingency as it would have appeared to his hundredth ancestor (about the time of King Solomon) would require for their expression forty-four figures of the decimal notation; yet this improbable event has certainly happened.

R. A. Fisher (1954)

L'an mil neuf cens nonante neuf sept mois
Du ciel viendra un grand Roy d'effrayeur
Ressusciter le grand Roy d'Angolmois
Avant après Mars régner par bonheur.

Nostradamus, Quatrain 72, 10th Cycle

Over the years I have made only rare and short-lived attempts to maintain a diary—probably a good thing, as the fragments that survive reveal a life largely lacking in excitement and drama. The most sustained attempt occupied the first six weeks of 1959, with the result that I know with some exactness how and when I came to hear the name of Fred Hoyle, one of the most important cosmologists of the twentieth century, but also notorious for his excursions into biological evolution. On January 22, 1959, I recorded that my religion teacher had been reading a lecture by Fred Hoyle, and that I disagreed strongly with it about relativity. It seems a curious thing to be reading in a religious knowledge class, but the teacher was making a contrast between what he saw as Hoyle's godless approach to science with a more spiritual one we had heard about the previous week in a recording by another cosmologist, Sir Bernard Lovell, famous at the time but now, unlike Hoyle, largely forgotten.

Although I remember the occasion independently of the diary, my memory is different from what I wrote, in that I can remember feeling contemptuous of Hoyle's suggestion that somewhere in the universe there was a team that could beat Australia at cricket, but I have no recollection of disagreeing with him about relativity; indeed, I am astonished that at that time I thought myself qualified to hold an opinion about relativity, especially one in strong disagreement with that of one of the world's experts. The remark about cricket teams elsewhere in the universe seems at first sight too absurd to be worth discussing, but it makes sense if the universe is infinitely large. In an infinitely large universe every possible event, no matter how improbable, is occurring, not just once but infinitely often.

The specific improbable event that interested Hoyle throughout much of his life was the appearance of life and its subsequent evolution into the diversity of life that we know today. He believed that both the initial appearance and the subsequent evolution were too improbable to have occurred in such a small place as the earth, and that the whole idea of evolution of complex life forms from simpler ones on earth was absurd. He believed we needed the vastly larger space of the entire universe to explain it. The problem with this argument is that the universe as it is described by most astronomers is not nearly large enough to be of much help. We are so conscious of how much bigger it is than ourselves that we easily forget how small it really is.

Although Hoyle himself sometimes referred (without explanation) to a "spatially infinite universe, a universe that ranges far beyond the largest telescopes," the usual estimate is that it contains about 10^{79} atoms (one followed by 79 zeroes), a trivial number compared with what we need if we are to be convinced of the necessity to assume an origin and evolution of life elsewhere. This many atoms may seem like quite a lot, but let us consider it in relation to more everyday objects. As most of the 10^{79} atoms are hydrogen atoms, it corresponds to a mass of the order of 10^{52} kg, about 10^{27} times heavier than the earth, which weighs about 10^{25} kg. By contrast, the common gut bacteria *Escherichia coli* have cells with a volume of about 10^{-15} liter, and hence a mass (taking the density to be about that of water) of about 10^{-15} kg. Thus the earth is about 10^{40} times bigger than a typical bacterial cell, and is vastly bigger on the scale of such a cell than the universe is on the scale of the earth, as illustrated schematically in Figure 11.1. Taking a different example, an adult human weighs of the order of 100 kg (most of us weigh less than that, but no matter, we are talking in terms of orders of magnitude here to make the arithmetic easy), and the sun weighs about 10^{30} kg, so it is about the same size in relation to the universe as a human being is in relation to the earth. Of course, the sun and the other stars are much too hot for life to have originated, evolved, or survived there, and the dark matter that is believed to account for most of the mass of the universe is much too cold (with an average temperature a few degrees above absolute zero). The earth, therefore, represents a larger proportion of the inhabitable part of the universe than these calculations suggest, and anyway a far larger proportion than most of us would guess without doing this simple arithmetic.

What does it matter if the universe is only 10^{27} times heavier than the earth, rather than, say, 10^{270} or 10^{2700} times heavier? Surely even 10^{27} is a very large factor, large enough to cope with the improbabilities inherent in the appearance

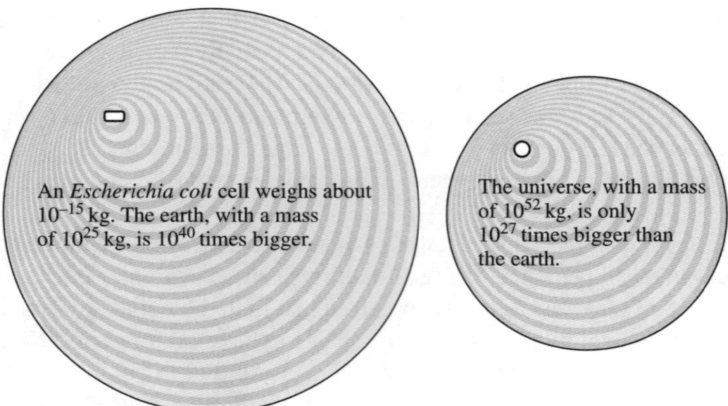

Fig. 11.1 Relative sizes of a bacterial cell, the earth, and the universe. The universe is much smaller on the scale of the earth than the earth is on the scale of an *E. coli* cell. The drawings are not, of course, drawn to a linear scale, but each shaded background circle represents a 100-fold increase in size

and evolution of proteins? Unfortunately for this argument, a factor like 10^{27} is utterly trivial in relation to the sort of numbers that Hoyle and others calculate in their efforts to prove that the biologists' view of the origin and evolution of life is untenable. Their argument is as follows: Let us take a "typical" protein with a sequence of 100 aminoacids, such as the protein cytochrome *c*, which does have about 100 aminoacids in many species, and is found in all the animals, plants, and fungi that have been checked. In the human, for example, it begins Gly-Asp-Val-Glu-Lys-Gly-Lys-Lys-Ile-Phe . . . , each group of three letters here being the three-letter symbol for one of the 20 aminoacids that are found in proteins. Suppose that the probability of having Gly (glycine) in the first position would be about one-twentieth, or 0.05, if the sequence were arranged at random, and that of having Asp (aspartate) at position 2 would likewise be 0.05, that of having Val (valine) at position 3 would again be 0.05, and so on. (These are suppositions rather than necessary facts because they involve assuming that all the possibilities are equally likely. In fact they are not, as some aminoacids are much commoner than others, and if this is taken into account the true probabilities come to about one-fourteenth, or 0.07, at each position. However, this makes very little difference to the argument, and I shall ignore it.)

If we further suppose that whatever aminoacid we have in one position has no effect on the probability of what we will find adjacent to it, then the probability of having both glycine at position 1 and aspartate at position 2 is 0.05×0.05, or $0.05^2 = 0.0025$, the probability of having Gly-Asp-Val as the first three is $0.05 \times 0.05 \times 0.05$, or $0.05^3 = 0.000125$, which is more convenient to write as 1.25×10^{-4}, and so on. By the time we have counted to 10 aminoacids we have $0.05 \times 0.05 \times 0.05 \times 0.05 \times 0.05 \times 0.05 \times 0.05 \times 0.05 \times 0.05 \times 0.05 = 0.05^{10} = 9.8 \times 10^{-14}$ as the probability that a random sequence will begin Gly-Asp-Val-Glu-Lys-Gly-Lys-Lys-Ile-Phe. . . . This is already very small, but this is just the beginning. By the time we get to 100 aminoacids we shall have $(9.8 \times 10^{-14})^{10}$, or 10^{-130}. Human cytochrome *c* actually has 104 aminoacids, not 100, and the extra 4 aminoacids bring the probability down to only 10^{-135}.

This is actually a very optimistic estimate of our chances of producing human cytochrome *c* by random shuffling of aminoacids, because it assumes that there is some way of doing the shuffling and stitching all the aminoacids together in the first place. That is in itself very improbable, and it is just the beginning of the difficulties: most enzymes are much bigger than cytochrome *c* (they have more than a hundred aminoacids) and we need more than one of them to make a living organism. A human cytochrome *c* molecule is not a human being! Even if we optimistically guess that we could have a viable organism with as few as 100 different proteins (remember from Chapter 3 that *Mycoplasma genitalium*, the simplest known modern organism, has around five times that many), with an average of 100 aminoacids each, and we skate over the difficulties of putting the right amounts of these proteins together at the right place, and protected from the environment by a suitable cell wall, we still find ourselves with a probability of about $10^{-13\,500}$, an almost inconceivably small number.

If we pursue this argument we arrive almost inevitably at the conclusion that we can "prove" that it is impossible for life to have originated on earth, and although that may seem like an argument that it originated somewhere else, that is not sustainable because, as we have seen, "somewhere else" is only 10^{27} times bigger than the earth, and improving our chances by a factor of 10^{27} is barely worth bothering with if we start with odds of one in $10^{13\,500}$ against us. How comforted would anyone feel to learn that some enterprise had a chance of one in $10^{13\,473}$ of success after thinking it was only one in $10^{13\,500}$? In any case, 10^{27} is very optimistic as an estimate of how much it helps to regard the whole universe as available for the origin of life to have occurred. Not only is most of the universe much too hot or too cold for life, but, in addition, the hypothesis that life originated elsewhere introduces a new difficulty that does not arise if life on earth originated on earth: we need a mechanism to transfer existing life forms to the earth in a viable form from wherever they originated, so whatever we calculate as the probability that life originated anywhere must be multiplied by the very low probability that such a mechanism exists.

There is one aspect of this analysis that we have not yet touched upon. Just as we tend to assume that the universe is vastly bigger than the earth unless we make the appropriate numerical comparison, so we also tend to think of the universe as being vastly older than the earth. If the universe is, say, $10^{13\,500}$ times older than the earth then maybe there has been so much time available for extremely improbable events to have occurred elsewhere in the universe that there is no longer any difficulty with postulating them. The problem with this argument is that the universe appears not to be nearly old enough. Although at the time when Hoyle first began to advance his probability arguments there was some support among other cosmologists for his theory of an infinitely old universe, this appears to have largely dissipated in the past 40 years, and the current view favors a universe that is perhaps three or four times older than the earth—about 14 billion years for the universe and about 4 billion years for the earth. It would be presumptuous for a biochemist to enter into this argument; suffice it to say there is no support for an infinitely old universe in the writings of authorities of the standing of Stephen Hawking, in *A Brief History of Time*, Steven Weinberg, in *Dreams of a Final Theory*, and Roger Penrose, in *The Emperor's New Mind*.

Thus not only can we apparently prove that life could not have originated on earth, we can prove almost as easily that it could not have originated anywhere. Yet this conclusion must be wrong: we are here to discuss it, and we *know* that life does exist on earth, so any proof otherwise must be flawed. It seems to me that there are only two possible ways out of the difficulty: either a divine intelligence planted life on earth (and possibly elsewhere) in a deliberate act, or the probability calculations outlined above are incorrect. No biologist would accept the counsel of despair implicit in invoking a divine intelligence, and recalling that back in 1959 my teacher of religious knowledge used him as an example of a godless scientist, I do not believe that Hoyle would have done so either; even theologians nowadays appear unenthusiastic about a creator who wound up the clock, defined the laws of physics and then left it running for 14 billion years without interference.

This leaves the second possibility, and virtually all modern biologists would agree that the explanation is that the probability calculations are at best inappropriate, and probably plain wrong as well. However, we cannot just leave it at that: we need to give some plausible reasons for thinking that the calculations may be wrong. Let us begin by noticing that they implicitly assumed that the particular 104-aminoacid sequence that cytochrome *c* has in the human is the only possible one it could have if it is to work. If, on the other hand, the particular sequence is one of, say, 10^{120} that would work just as well, then the chance of getting a working cytochrome *c* by shuffling aminoacids ceases to be a hopelessly improbable endeavor. We have no justification for putting the number as high as that, but we certainly have reason to believe that it is bigger than one, that is, that the sequence we have is not the only possible one. First, although not only humans but also those other mammals that have been studied have cytochrome *c* sequences that begin Gly-Asp-Val-Glu-Lys-Gly-Lys-Lys-Ile-Phe . . . , the mammals do vary from one another in the remaining 94 locations, and some birds start with Gly-Asp-Ile-Glu-Lys-Gly-Lys-Lys-Ile-Phe . . . , the bullfrog begins in the same way as the mammals, but the tuna has Gly-Asp-Val-Ala-Lys-Gly-Lys-Lys-Ile-Phe . . . . The aminoacid Ile (isoleucine) is rather similar in its properties to Val (valine), and replacing valine by isoleucine is usually regarded as a "conservative" change. On the other hand, Ala (alanine) is rather different from Glu (glutamate), so this is not a conservative change.

Examination of the whole range of known cytochrome *c* sequences reveals at least some variation at the majority of loci, and they are not all of the same length: even among higher animals there is some variation, so that frogs and fish have 103 aminoacids instead of 104; most plants have somewhat longer sequences. All of these proteins have very similar chemical and physical properties, so it is certainly not the case that a unique sequence is needed to do the job. On the other hand, there is no suggestion that a vast number of different sequences could do it either. Among other kinds of proteins there are some that are more conservative than cytochrome *c* (such as histone IV, a protein associated with the genetic material, DNA), which shows almost no variation over the entire animal and plant kingdoms. There are also many that are much more variable, with barely detectable sequence similarity in comparisons between distantly related animals.

Unfortunately none of this tells us whether even the most conservative structure, such as that of histone IV, is really the only one possible, or whether it is just that having once chosen one of many possible structures organisms are now "locked" into it because although others still exist they cannot be reached by any reasonable series of mutations. Consider a bee living on a small island 20 km from the mainland. Such a bee might well conclude from its forays of a few kilometers from its hive that the island where it lived was the only place in the entire universe where life was possible, but it would be not only wrong but grossly wrong: a huge number of other places exist where bees can live, but they just happen to be inaccessible.

In Chapter 3, I discussed some of the knowledge that we have of how much variability in protein sequences organisms can stand. Further suggestions that a wide variety of sequences might be capable of fulfilling the role of any given protein come from studies of lysozyme, an enzyme that is widely distributed and acts to break down the cell walls of certain bacteria. It was famous for a while as a sort of "pre-penicillin," as it was discovered by Alexander Fleming before he discovered penicillin. Noticing its capacity to destroy some bacteria, he hoped it might have therapeutic value as an antibiotic. In fact it proved not very useful for that, but its discovery may still have prepared Fleming's mind for the idea that much better antibiotics, such as penicillin, might exist. For the present discussion, the point is that lysozyme's bacteriolytic activity can be mimicked quite well with random copolymers of aminoacids. For example, a "protein" with a sequence consisting of the two aminoacids glutamate and phenylalanine in a random order has about 3% of the bacteriolytic activity of lysozyme.

Honesty compels me to admit that there are some difficulties with this result. As mentioned already, lysozyme is not a particularly good bactericide, as it is inactive against the more dangerous bacteria. In any case it is found in places that are not very susceptible to bacterial infection (like bone marrow), and it is not found in some places where its bactericidal properties would be more useful. So, although this is the conventional view of lysozyme's function, it is possible that its real function is something different that has not yet been identified. In this case, the ability of random proteins to mimic its bacteriolytic activity is irrelevant. In an emergency you could use the pages of this book to kindle a fire, and they would do the job quite well; you might then find that some dry leaves from your garden would make an adequate alternative, but that would tell you nothing about the fitness of the book for its primary purpose.

Two other enzymes perhaps offer a different sort of evidence. Carbonic anhydrase catalyzes the dissociation of carbon dioxide from the state that exists when it is dissolved in water. I mentioned it in Chapter 1 as an example of an enzyme that catalyzes a reaction that occurs very fast with no catalyst at all, noting that there are some contexts, for example for releasing carbon dioxide gas in the lungs for exhalation, where completing the reaction in a few seconds is not fast enough. Remember that we (humans) breathe about 12 times/min, and so the average time available for the carbon dioxide bound to the hemoglobin that arrives in the human lung to be exchanged for oxygen is about 5 s. The rate at which dissolved carbon dioxide can be dehydrated if there is no catalyst is thus insufficient. Carbonic anhydrase is found in many different organisms in addition to mammals, and it exists in three known classes with no detectable structural similarity. In other words, there are three entirely different kinds of proteins with no resemblance between them if you just compare their structures, but which catalyze the same process and fulfill the same biological function.

Another enzyme, superoxide dismutase, is essential for preventing some highly toxic side effects of the use of oxygen in the energy-harnessing apparatus of many organisms and is found in two entirely different classes with no detectable structural resemblance. As a digression—incidental so far as this chapter is concerned, but an interesting example of the sort of things that can happen in evolution—the ponyfish (*Leiognathus splendens*) is a luminous fish that emits light by means of a light organ in which lives a light-emitting bacterium called *Photobacter leiognathi*. Both fish and bacterium have superoxide dismutases, and that of the fish is a typical member of the class that other vertebrates have. The bacterium, however, has two kinds of superoxide dismutase, not only the class expected in bacteria but also a second enzyme that is typically fish-like. It is quite unlike the enzymes found in other bacteria, but at the same time not so similar to the enzyme found in its host that we could explain its presence by contamination of the bacterial samples with fish tissues. It seems hard to escape the conclusion that a "horizontal transfer" of DNA took place during the long period of association between the two species, that is, at some moment a bacterial ancestor incorporated some fish DNA in its genome.

Comparisons between the sequences of enzymes fulfilling the same functions in different species give at least a minimum estimate of how many *related* sequences are capable of doing any given task. In most cases (histone IV being an outstanding exception) we find at least one or two differences between the sequences found even in closely related organisms. So the number of sequences possible for an enzyme that occurs in all species should be at least as large as the number of species, making millions of sequences (even if we admit only species that exist on earth today, but many more if we include extinct species). More information comes from experiments that have been done to "shuffle" the DNA in particular genes, for example by Willem Stemmer and colleagues. The idea is to take a set of genes from different strains of a bacterium that code for forms of an enzyme that are all different, but which are all capable of catalyzing the same reaction. If the genes are broken into fragments, mixed together, and then reassembled in a way that allows each reassembled gene to contain a random collection of fragments from different sources, the result is a large collection of genes coding for related but different protein sequences. When these genes are expressed as proteins some of them, of course, turn out to be nonfunctional, but, more interestingly, large numbers of novel proteins do act as catalysts, some of them better than the natural ones. So the number of related sequences capable of fulfilling any given function is typically of the order of millions or more.

Unfortunately, these experiments tell us nothing about how many different *unrelated* sequences would do the job, and we have almost no way to generalize from the observations with carbonic anhydrase and superoxide dismutase to decide how much structural tolerance there may have been in the primitive enzymes that existed at the origin of life. We can be quite certain, however, that there was more tolerance than there is for any organism today, because the first living organism had no competitors. A rival to *E. coli* appearing today in the human gut would rapidly disappear if it needed a day to reproduce itself, because in that day an *E. coli* cell (with a doubling time of about 20 min) could have been replaced by more than 10 million copies of itself. No matter if in some respects the rival had a more efficient metabolism than *E. coli*; no matter if each cell secreted a powerful toxin capable of killing a thousand *E. coli* cells; it would still be outgrown and would never be detected.

These concerns would not have mattered to the first organism; there was no race to reproduce because there were no competitors. So we can guess that the first organisms could survive with many fewer enzymes than any modern organism needs, and that the individual enzymes could have been far less efficient at fulfilling their functions than their modern counterparts, but beyond that rather vague statement we cannot easily go. I think that few biologists would claim that the origin of life is a solved problem, or even a problem that will be easy to solve to everyone's satisfaction in the next century or so. But few would go to the other extreme either and agree with the creationists that the probabilities against the origin of life are so enormous that it will never be solved without invoking a divine creator, or, as Hoyle and his supporters would prefer, seeding of the earth with bacteria grown elsewhere in the universe.

Extremely unlikely events happen every day, and we can find examples of things that have certainly happened but which have calculated probabilities so small that we could easily be tempted to regard them as impossible. Consider the quotation from R. A. Fisher at the beginning of this chapter. He assesses the probability that a man living at the time of King Solomon would have a descendant in the direct male line living one hundred generations later as being of the order of one in 10^{44}, though, as he adds, this improbable event has certainly happened. In case this last point is not perhaps completely obvious, let us spell it out. Any adult man living today has or had a father, so that father had at least one son who survived to maturity. His father likewise had at least one son who survived to maturity, and so on backwards for one hundred generations.

Why stop at a hundred, however? Humans are now thought to have become differentiated from chimpanzees and gorillas around 3–5 million years ago, so let us imagine a protohuman sitting by his camp fire on an African savannah at the time of Lucy, around 3 million years ago, and pondering the likelihood that he would in the twentieth century have a distinguished cosmologist by the name of Fred Hoyle as his descendant through the direct male line. If he was as good at estimating chances as Fisher was, he would arrive at an almost inconceivably small number, given that at least a hundred thousand generations must have passed in 3 million years. He would probably conclude that the thing was impossible. Yet, if not for him then for one of his contemporary protohumans, this improbable event has certainly happened.

Fisher's point here is that we must refrain from getting carried away by improbabilities viewed after the event. Probability calculations in relation to future and unknown events may be very helpful, and are among the things that separates good bridge players from the rest, but a posteriori probability calculations are rarely very illuminating.

I claimed a moment ago that extremely unlikely events happen every day. How could I possibly justify such an extravagant claim? Before doing so, I shall put it even more extravagantly: every day, some person on earth has an experience that has odds of worse than one in a billion against its occurrence at that time and in that place. Surprising though it may seem, there is no question that this is true. There are many more than a billion people in the world (there are more than that in India alone), and even if we make the conservative estimate that each of them has just one "experience" each day that is in some sense particular to that person, that still makes at least a billion separate experiences each day. Of these, 95% will not be "significant" in the sense commonly used for statistical tests of experimental observations. This means that 95% will fail to satisfy tests to detect observations outside the range of what is expected 95% of the time (assuming that the tests have been appropriately designed and correctly done, conditions that are often, unfortunately, unfulfilled). Likewise 99% will fail to be "highly significant," where this has a meaning that you can easily guess from the way I defined "significant."

However, this leaves 1% of highly significant experiences, or 10 million of them every single day, even on a very conservative estimate of how many experiences a person has each day. By an obvious extension we see that every day more than a thousand people have experiences that are expected only once in a million trials, and on an average day one person will have a one-in-a-billion experience. Given that there are quite a lot more than a billion people in the world, and each of them has more than one experience each day, it is surely fair to multiply by at least 10, so that more than 10 people have one-in-a-billion experiences each day, or one person has a one-in-ten-billion experience. But even this is too cautious, as we are not mayflies and most of us live much longer than a day. In the biblical threescore years and ten there are more than 25 000 days, so it will be fair to say that in the course of a typical lifetime someone somewhere will experience something that is not expected to occur more often than once in 2.5×10^{15} trials.

To put this in a more familiar perspective, if everyone on earth was dealt an average of one bridge hand per day, all of them with proper shuffling and no cheating, about once every 2 years someone would be dealt all 13 hearts. Or, for those that remember the Rubik's cube craze of the early 1980s, if everyone on earth spent at least 3 h/day making random moves on shuffled Rubik's cubes, making about one move every second (if you do remember the early 1980s you may feel that it did indeed seem like that at the height of the craze), we should expect a report of chance success about every century. This is not a very high frequency, to be sure, but it is much more than the "impossible" odds of one in 4×10^{19} might lead you to expect.

Coming back to evolution, and bearing in mind that something of the order of a thousand million years were probably available in the prebiotic period before anything very interesting happened at the origin of life, and bearing in mind that we are not talking about a mere billion conscious people having experiences, but a much larger number of localities on earth where haphazard "experiments" were being done by mixing chemicals in arbitrary ways determined by variations in the weather, tides, volcanic and meteorite activity, etc., we can imagine that huge numbers of combinations were tried out over a vast space of time. Even if the overwhelming majority of these experiments yielded nothing very interesting in the way of results, some extremely unlikely things could happen in the time available.

It remains to explain why this chapter is prefaced with a quotation from Nostradamus. It is there to contradict another claim made by Hoyle in one of the last articles that he wrote to justify his view of the impossibility of an independent origin of life on earth. As I mentioned, he made numerous references to "bee-dancing" to characterize the fuzzy thinking of biologists, which he contrasted with the "real science" practiced by people like himself, and exemplified by the accurate prediction of the total eclipse of the sun on the August 11, 1999. This was an unfortunate example to have given, because, not only could the eclipse have been predicted many centuries before the tools of "real science" were developed, it was in fact predicted by Nostradamus about a century before Isaac Newton and around four centuries before Albert Einstein.

There are two objections you might want to make about this "prediction." First, it refers to the seventh month of 1999, but August is not the seventh month. I could dismiss this as a minor inaccuracy, but in fact it is not even that, because according to the Julian calendar in use at the time of Nostradamus, the eclipse was indeed in the seventh month.

Second, you may object that although the date is very clear the rest of the quatrain is very vague. A reference to the sky, certainly, but what have a king of terror and the King of Angoumois to do with eclipses? We can answer this with another question: what event in July or August 1999, other than the eclipse, could a rational person conceivably regard as being accurately predictable in the sixteenth century? Clearly nothing, so if it is a prediction at all it can only refer to the eclipse. Moreover, although most of Nostradamus's dates are as vague as the rest of his text, the few that definitely refer to events in what was then the future can all be associated with astronomical events that would have been predictable by the methods available to him. In the circumstances it would be perverse to interpret it as a reference to anything but the eclipse of August 11, 1999.

One difficulty with disagreeing with physicists—even about biology—is that there is a hierarchy in science whereby lesser scientists are supposed to accept the opinions of physicists as if they were holy writ. This applies as much to popular books and magazines as it does to serious scientific journals. A book questioning the validity of natural selection stands a good chance of being published if it comes from the pen of a physicist or an astronomer, and of being taken seriously by at least some of its readers. A biologist who wrote a book questioning Copernicus's heliocentric view of the solar system would be very unlikely to find a publisher. If such a book did appear it would make a laughing stock of the author, and might well provoke articles in serious newspapers about the lamentably low level of culture among biologists.

Nonetheless, physicists have been wrong before in their arguments with biologists, and will doubtless be wrong again. Hoyle did not claim originality for his view that life could not have originated on earth, but traced it back to Lord Kelvin (at the time known as William Thomson) in the nineteenth century, who asserted that "we all confidently believe that there are . . . many worlds of life besides our own." This is, of course, the same Lord Kelvin who argued that Charles Darwin could not possibly be right in his theory of natural selection, and was equally confident that the earth could not be more than 400 million years old, and might even be as young as 25 million years old, and thus much too young for natural selection to have operated in the way Darwin needed. No physicist today thinks the earth is only 25 million years old! Kelvin is perhaps too easy a target (though he was Hoyle's choice, not mine), and among his other pronouncements from on high we may note his views that "heavier-than-air flying machines are impossible" and that "radio has no future."

All of this raises the question of why some physicists appear to take such a naive view of biology, even though they clearly are not naive about their own subject. In his interesting book *Darwin's Dangerous Idea*, Daniel Dennett suggests that what physicists miss in biological theory is a set of laws that allow the properties and states of systems to be expressed in simple (or not so simple) mathematical equations. But natural selection is not a law in this sense; rather it describes a procedure, what computer scientists would call an algorithm, that directs how systems change over long periods of time. Approached with understanding, natural selection allows a vast array of individual biological observations to be brought together in one rational framework. Without natural selection they are just a huge collection of unrelated facts, and I close this chapter by recalling the words of Theodosius Dobzhansky quoted already in Chapter 4: "Nothing in biology makes sense except in the light of evolution."

12
Genomic Drift

The jigsaw-like fit between Africa and South America has been obvious to everyone who has looked at a world map since sufficiently accurate maps became available. Francis Bacon commented on it in 1620, as did many others long before 1912, when Alfred Wegener proposed his theory of continental drift to explain it. Nonetheless, until the 1960s, the standard view of geophysicists was that the fit was pure coincidence, and they heaped on Wegener's ideas the kind of scorn that is today reserved by virologists for the notion that human immunodeficiency virus may not be responsible for the collection of diseases known as AIDS.

The standard explanation of why continental drift was not taken seriously by geophysicists until the 1960s was that before the development of plate tectonics there was no plausible mechanism to explain how continents could move. True enough, but that did not prevent the same geophysicists from accepting even more fantastical explanations of why geological features on one side of the ocean match those on the other, and why the fauna of Africa and South America resemble one another in some respects. Both there, and in an embarrassingly large number of other parts of the world, land bridges were supposed to have existed in the past but to have sunk beneath the seas later on. Despite the lack of explanation of how dense matter had once been able to float, or, if you prefer, how buoyant material was later able to sink, land bridges were regarded as more acceptable than drifting continents. The fact that Wegener had no recognized expertise in geophysics no doubt had something to do with it as well: he was the son of an evangelical minister, he was trained in astronomy, and his practical experience was in meteorology. Nonetheless, he was essentially right.

There are some similarities (but also some important differences) between the history of continental drift and that of the causes of cancer. It has been known for more than a century that there is a close association between cancer and a chromosomal abnormality known as *aneuploidy*, but most experts today consider this little more than a coincidence, or that the chromosomal abnormality is more of a consequence than a cause of the cancer.

To understand what is meant by aneuploidy we need to start from the generalization that most animals and many plants are diploid, as discussed in

Chapter 9. This means that in normal individuals all of the chromosomes other than the sex chromosomes come in pairs, and even the sex chromosomes are paired in one of the two sexes. In humans, for example, there are 22 pairs of autosomal (i.e. nonsex) chromosomes, together with two sex chromosomes, which form a pair in women (XX) but not in men (XY), making 46 chromosomes in total. As always in biology, it is more complicated than that, because even though women have two X chromosomes only one of these is expressed in any one cell. Thus, so far as expression is concerned, only the autosomal chromosomes are paired, and there is no difference between men and women in this respect.

Different species have different numbers of chromosomes, and in Chinese hamsters, an example whose relevance will become clear later on, there are 10 pairs of autosomal chromosomes and one pair of sex chromosomes, making 22 in total. If no mistakes are made during cell division, the counts are preserved after each such division, but this is a highly complicated and still only partially understood process, and mistakes sometimes do occur; as a result a daughter cell may contain one or three examples of an autosomal chromosome instead of the expected two. When this occurs in the germ line the result may be an individual with an abnormal number of chromosomes in every cell, resulting, as noted briefly at the end of Chapter 1, in conditions such as Down syndrome and other trisomies. Errors also occur in somatic cell division, usually with less-serious results as only a minority of cells are affected. In either case—in individual cells, or in whole organisms—the occurrence of an abnormal number of chromosomes is called aneuploidy.

As long ago as 1890 David von Hansemann noticed that the human cancer cells that he examined were aneuploid, and he suggested that the aneuploidy caused the cancer. Later on, Theodor Boveri developed this idea into a general theory of cancer, and for many years this was considered to be a respectable hypothesis. Even though the connection between effect and supposed cause remained obscure, many facts were known about cancer that were consistent with aneuploidy as its cause. First, cancer cells have abnormal rates of growth, hardly surprising if they have a chromosomal composition different from what the cell-division machinery "expects." In popular accounts this abnormal rate of growth is usually understood as abnormally *fast* growth, but that is an error, as many cancers grow abnormally slowly; what is general is that they do not grow at the same rates as the normal cells that gave rise to them. Second, they show extreme genetic instability, so that in each generation the daughter cells are genetically different from their parent cells, in marked contrast to the genetic identity (or near-identity) that persists through many cell divisions in normal cells. Third, cancers display metabolic defects, with abnormally high metabolic activity, though they may sometimes lack some metabolic pathways completely. All of these and other well-established characteristics of cancer can be understood in terms of Boveri's hypothesis, as we shall see later on, but that did not prevent it from being largely discarded and forgotten about in the 1970s after the discovery that certain genes, known as *oncogenes*, are characteristic of certain kinds of cancer.

The first of these stemmed from Peyton Rous's discovery in 1910 that a type of cancer common in chickens was caused by a virus and could be transmitted artificially from one chicken to another. In 1966 he was awarded the Nobel Prize for this discovery made more than half a century earlier. Not long after this, Peter Duesberg discovered the oncogene responsible for the infectious character of this cancer. He found that chickens infected with an RNA virus containing such a gene developed cancer, but did not develop cancer when infected by a virus from which the oncogene had been removed.

This was the first oncogene to be characterized, and it certainly did not establish that *all* cancers are caused by oncogenes. However, the idea that oncogenes were needed for cancer proved so appealing that within a few years it became the dominant hypothesis, and largely supplanted the aneuploidy hypothesis in the minds of most cancer researchers. An exception, ironically enough, was Duesberg, who has continued to regard viral oncogenes as a rarity, and has gathered accumulating evidence that Boveri was right all along.

Everyone agrees, of course, that cancer cells are aneuploid; they could hardly do otherwise, as the evidence is overwhelming. The disagreement is over whether aneuploidy is the cause or a consequence of the cancer. The standard view is the latter, as it is claimed that no mechanism exists to explain why aneuploidy should result in cancer; in any case "everyone knows" that genetic mutations cause cancer and that mutagens, or substances in the environment that generate mutations, are also *carcinogens*, or cancer agents. There are, however, some major exceptions: the carcinogenic effect of asbestos fibers is now so generally accepted that the use of asbestos for insulating buildings is banned throughout the developed world, yet asbestos fibers are chemically inert and efforts to show that they cause mutations have completely failed. What is more, the capacity of asbestos to prevent normal cell division, and hence to cause aneuploidy, is only too clear: photographs taken with microscopes show cells with embedded asbestos fibers making purely physical barriers to the proper organization of chromosomes necessary for normal cell division. There is no more need to invoke chemistry to explain this than there would be to invoke defective steering wheels to explain the effect on the circulation of traffic of a tree fallen across a road. This is much like the argument I have used in Chapter 7: let us reserve our sophisticated explanations for the things that cannot be explained more simply, and not waste them on things where a simple explanation accounts for all the facts.

Exceptions like asbestos can perhaps be set on one side. After all, as I have also insisted in Chapter 7, biological phenomena are so extremely complicated that we should not expect too close an agreement between prediction and results, and the sort of perfect correlation that would delight a physicist should arouse suspicion in a biologist. So even the best of biological theories usually have to contend with one or two inconsistencies that have to be left to future biologists to clarify. There remains, however, the question of timing. If aneuploidy is the consequence of cancer, then we should expect to find some cells that are clearly cancerous but which have not yet become aneuploid, and, of course, vice versa: if aneuploidy is the cause we should expect to find noncancerous aneuploid cells. The latter observation is easy to make: after all, any individual with Down syndrome or another trisomy is an entire organism of 10^{14} aneuploid but noncancerous cells. Moreover, although such trisomic individuals are not necessarily cancerous, they do have a high probability of developing cancer. A person with Down syndrome, for example, has about 30 times the average probability of developing cancer.

What about the converse observation: can we find cancerous cells that are not aneuploid? Duesberg and his collaborators studied this question by examining the effects of the carcinogenic chemical dimethylbenzanthracene on Chinese hamster embryo cells. As noted above, a normal diploid cell of this animal has 22 chromosomes, but 3 weeks after treatment the commonest chromosome number was 23, many cells having other numbers ranging from 20 to 25, and a few having numbers about double those in this range, 44 or 46 (Figure 12.1). In the following week the counts below 22 or a little above 23 decreased, but many other values between 28 and 46 appeared. All of this, incidentally, occurred months before there were any signs of cancer. In untreated cells used as a control the majority of cells had 22 chromosomes,

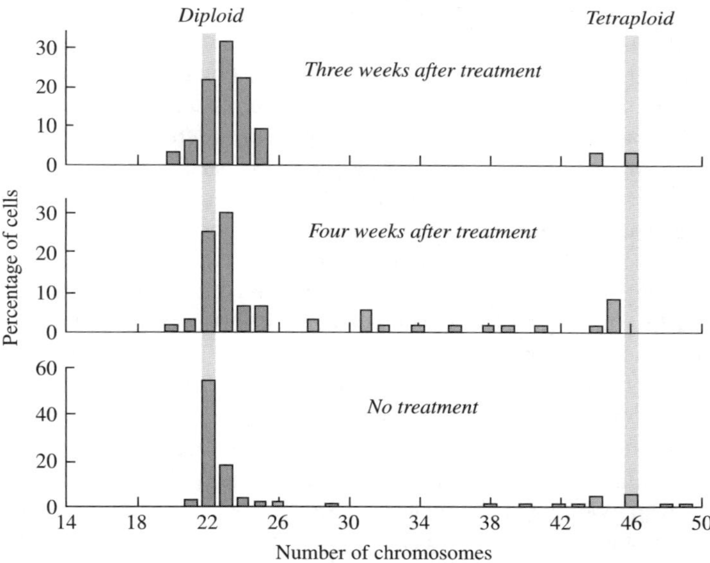

Fig. 12.1 Development of aneuploidy in Chinese hamster cells. Three weeks after treatment of cells with a carcinogenic chemical (top) only about one-fifth of the cells have the expected number of chromosomes, 22. Many cells have somewhat fewer or somewhat more, and a few have approximately double the expected number, so they are tetraploid rather than diploid. One week later (middle) there are numerous cells with chromosome counts intermediate between the diploid and tetraploid values. Aneuploidy is quite common even in normal healthy cells (bottom), and only a little more than half of the cells are strictly euploid.

though about a quarter had 23, and other numbers, as high as 49, occurred with detectable frequency.

These observations are sufficiently complicated that it would be rash to propose a simple explanation. Nonetheless, several points appear clear. First, aneuploidy is by no means absent from the untreated cells, and more generally, a substantial proportion of cells are aneuploid in individuals who do not develop cancer. Thus cancer cannot be a *necessary* consequence of aneuploidy, and the normal organism must be able to tolerate a moderate degree of aneuploidy. The simplest explanation is that although the aneuploid cells in a normal population are defective and likely to die out after a small number of cell divisions, this does not matter very much as long as they remain rare, and as long as they can be replaced by normal ("euploid") cells as fast as they die out.

In the precancerous population of Chinese hamster cells, the increase in complication between the third and fourth weeks can perhaps be interpreted as evidence of the efforts of the population to correct a problem that is beyond correction. As these efforts can be seen to be making the problem worse, it should not be too surprising if total failure, in the form of cancer, appears some weeks later. Finally, we should note in passing that a count of 22 does not guarantee that a cell is euploid, with a normal set of chromosomes: it could have three or more examples of some and only one or zero of others.

All of this may seem to be taking us rather far from the main themes of this book, but a few years ago Duesberg and his collaborator David Rasnick realized that metabolic control analysis could provide the key to explaining how aneuploidy could lead to cancer. As we have seen in Chapters 8 and 9, increasing or decreasing a gene dose by 50% usually produces no easily observable effect at all. Even complete elimination of a gene often has much less effect than one would guess. This came as a considerable surprise to people who developed techniques for producing mice that lacked selected genes: the characteristics of such "knock-out mice," as they are called, were supposed to reveal the functions of the genes that had been suppressed. Distressingly often, however, such mice have proved to be healthy. Similar experiments can be done on a much greater scale in organisms like yeast, and it turns out that around 80% of yeast genes are "silent": if such a gene is eliminated, the yeast continues to grow, in many cases at a normal rate. Does this mean that most genes are unnecessary and that the organism could manage perfectly well without them? Well no, it just means that organisms have back-up mechanisms to handle problems that occur when components fail, and that some genes have functions that are needed in extreme conditions and not all the time. A well-fed laboratory mouse experiences a more comfortable environment than a wild mouse in the field; yeast cells growing in a thermostated fermenter encounter fewer stresses than cells living on the walls of a brewery.

However, the early stages of aneuploidy are more likely to involve changes of 50% in gene doses than complete elimination of genes. This is because any reasonable model of how aneuploidy arises will attribute it to a malfunctioning of cell division that leads to one or three examples of a given chromosome instead of the normal two. As we have seen in Chapter 9, the typical effect of changing any one enzyme level by 50% will be no visible effect, but in aneuploidy we are not concerned with "any one enzyme" but with all the enzymes that are encoded by the chromosome affected, and a large number of individually negligible effects can add up to a substantial total. This, after all, is just what the flux summation theorem says: every one of the flux control coefficients can be individually negligible, but together they account for the whole of the control of the flux.

So we need to consider the effect on a generalized flux representing the metabolic activity of the cell of changing many of the enzyme concentrations by 50%. If all of them are changed by this amount then the expected result is simple: any flux will also change by 50%, because no matter how many enzymes there are, their flux control coefficients add up to 1. However, an error in cell division will typically produce aneuploidy in just one chromosome: in Down syndrome, for example, there are three examples of chromosome 21 instead of two. Chromosome 21 is among the smallest human chromosomes, however, and the normal pair account for much less than one twenty-third (4.3%) of the human genome, about 1.8%, in fact. As even the smallest chromosomes carry many genes, however, it is not likely to produce a severe error if we assume that the average flux control coefficient of a gene product of chromosome 21 is approximately the same as the average for all the gene products. So we can estimate the metabolic effect of Down syndrome as that of increasing the activity of 1.8% of the enzymes by 50%, which comes to about 0.9%.

Actually we have assumed here that flux control coefficients remain constant when the enzyme activities concerned change, but, as noted in Chapter 8, this is too simple: a flux control coefficient normally decreases when the activity of the enzyme increases, and vice versa. In their analysis, therefore, Rasnick and Duesberg used a more realistic model that allowed for the changes in flux control coefficients when enzyme activities change, while remaining simplified in some other respects. With this model the metabolic effect in Down syndrome is calculated as 0.6%—smaller than 0.9%, certainly, but not so much smaller that the naive analysis is totally meaningless, especially as 0.9% is small enough to make the point: overexpressing the whole of chromosome 21 by 50% is expected to produce a barely detectable effect on any flux. This is, of course, consistent with the observation that Down syndrome does not produce gross metabolic perturbations, and, more important, it emphasizes that, even when many enzyme activities are changed, it is very difficult to produce large changes in metabolic rates.

In a Chinese hamster an average pair of chromosomes accounts for one-eleventh of the genes in the genome. But suppose the initial error in cell division affects one of the largest pairs so that, say, 20% of enzyme activities increase by 50%. In this case, the naive calculation ignoring changes in flux control coefficients can be calculated in the head, and is half of 20%, or 10%. The more realistic calculation gives about 7%, which is not negligible, certainly, but is still far from the gross metabolic abnormalities found in cancer. Thus a single appearance of aneuploidy as a consequence of an error in cell division is not sufficient by itself to produce cancer. However, as we have seen, aneuploidy beyond the normal range of variation is typically detectable in Chinese hamster cells a few weeks after treatment with a carcinogenic chemical, whereas cancer is only detected months after that. Moreover, the changes seen between the third and fourth week after treatment indicate that aneuploid cells become increasingly aneuploid with each cell division, increasingly abnormal from a genetic point of view. Thus each cell division in the precancerous stage produces a new generation of cells with a greater degree of genetic imbalance, a greater degree of aneuploidy, a greater degree of metabolic abnormality, and a higher risk of aneuploidy-caused cell death, until we arrive at a fully cancerous stage. Rasnick and Duesberg estimate that the problem gets beyond correction when more than about one-third of the cells are aneuploid.

What of the pursuit of perfection in all this? Appearance of cancer, followed by death, is hardly a progress toward perfection as we would usually understand it. However, that is to see the cancer from the point of view of the host, whereas we need to examine it from the point of view of the cancer cells. A cancer is not part of the host, as it has a different genome, and is not even of the same species, as it cannot interbreed with the host species and, again, it has an incompatible genome. Moreover, it dies with the host, and so it leaves no descendants. Thus a cancer is a parasitic species that lives out its entire evolutionary history from speciation to extinction during a fraction of the lifespan of its host. During that time, it passes through various generations and experiences natural selection like any other species, cells that are better able to survive in the peculiar environment of the cancer leaving more descendants than others.

Recognition of aneuploidy as the trigger that leads to cancer does not immediately suggest any particular change in the way cancer should be treated: once a patient has cancer, it is immaterial whether the original cause was an aneuploid cell division or some effect of an oncogene. On the other hand, the proper treatment of a disease is not the only medically important consideration. There is also the question of prevention, and with any disease understanding the causes helps to prevent its occurrence. So even though the analysis of metabolic control may not lead to cures for cancer, it may still lead to fewer cancer patients in the first place.

This is not the only way of applying the ideas of metabolic regulation to cancer, and improved treatment of cancer patients may follow from an analysis being developed by Marta Cascante in Barcelona. She has long been interested in metabolic regulation, and, in collaboration with colleagues in Los Angeles, she has been studying how current therapies may in some cases be based on faulty notions of metabolic control, and hence misconceived. Cancer cells typically devote a greater part of their activity than normal cells do to the production of nucleic acids, and, as we saw in Chapter 4, these include the five-carbon sugar ribose as an essential component. Cancer cells thus typically need more ribose than healthy cells, and, as we also saw in Chapter 4, they can get this by using the pentose phosphate pathway as a route for converting glucose and other six-carbon sugars into ribose.

Cascante therefore considered that the pentose phosphate pathway ought to be a good target for anti-cancer drugs, and accordingly examined which enzymes would be best to try to inhibit. As discussed in Chapter 8, most enzymes in any pathway have only a small influence on the flow of metabolites through the pathway, and as a result the effect of inhibiting them will be very slight unless the inhibition is virtually total. From this sort of analysis, Cascante identified the enzyme transketolase as a good target for inhibition, and also noted that the vitamin B_1 deficiency typical of cancer patients is directly related to the same enzyme. In biochemistry this vitamin is called thiamine, and it is necessary for the action of various enzymes, including transketolase. The high activity of the pentose phosphate pathway in cancer cells causes these cells to sequester much of the thiamine that would otherwise be used by the healthy cells.

The usual medical view is that thiamine deficiency in cancer is a problem that needs to be corrected, and the diet of cancer patients is accordingly supplemented with thiamine. Over a long period, of course, thiamine deficiency does create serious medical problems, being responsible for the disease beri beri, which was once very widespread. The disease does not develop overnight, however, and in the short term, during treatment of cancer, the cancer cells may well need the thiamine more than the healthy cells do, so adding it to the diet may do more harm than good.

The study of cancer is relevant to the main themes of this book not only as an illustration of how the rather abstract ideas of metabolic regulation developed in Chapters 8–10 can be brought to bear on a major medical problem, but also as part of the study of what life is. Healthy cells maintain a property known as *organizational invariance*, meaning that they maintain their identities as particular kinds of cells, or, if they are at the stage of development where the different kinds of cells in one individual organism are coming to be different from one another, they change their identities in a preprogrammed and orderly way. Cancer cells do not have this property, because each generation in the development of a cancer is different from the previous one, and within one generation the cells are heterogeneous, with unclear identities. Cancers always die with their hosts (unless artificially maintained as "cell lines" in the laboratory), but, even if they did not, their failure to maintain their organization would condemn them to death.

Cell lines are usually quite difficult to grow in the laboratory, though some rare exceptions, like a famous variety called *HeLa cells*, have continued to thrive more than half a century since they were first isolated in the 1940s from a cancer patient whose names began He . . . La . . . . These cells can be considered to have achieved a kind of immortality, and have presumably hit upon a structure that allows them organizational invariance.

In these closing remarks I have touched on a very difficult subject, much more difficult than anything else in this book, and I shall not develop it any further here, beyond indicating briefly what is difficult about it. The essential problem is that all of the artificial systems that we know and understand depend on external agencies to keep them running. If a machine in a factory breaks down or wears out, it cannot repair or replace itself but needs some other machine, or some other factory, to do so. Of course, a large and sophisticated factory can include machines to maintain the primary machines, but that does not solve the problem, because these secondary machines are also subject to wear and tear, and need to be repaired or replaced themselves. Any simple-minded attempt to solve the problem thus leads to an infinite regress, in which each new step raises more problems than it resolves.

The theoretical biologist Robert Rosen devoted all of his career to an attempt to close this circle, in other words to describe how an organism could achieve organizational invariance without falling into an infinite regress. His book *Life Itself* leaves biology with a major intellectual challenge that has not yet been taken up very seriously. Understanding organizational invariance will be essential for understanding what life is, and constitutes a major task for the twenty-first century.

Further Reading

As this book is intended for nonspecialists I have not thought it useful to give detailed references to the scientific papers. This is therefore a list of the books, mostly also intended for the general reader, that are mentioned in the text.

Ernest Baldwin (1964) *An Introduction to Comparative Biochemistry*, 4th edn, Cambridge University Press, Cambridge.

Francis Crick (1988) *What Mad Pursuit*, Weidenfeld and Nicolson, London.

Richard Dawkins (1981) *The Extended Phenotype*, Oxford University Press, Oxford. [Although this is directed more toward professional biologists than Dawkins's other books, it is readily accessible to the general reader.]

Richard Dawkins (1989) *The Selfish Gene*, 2nd edn, Oxford University Press, Oxford.

Richard Dawkins (1990) *The Blind Watchmaker*, Penguin Books, London.

Daniel C. Dennett (1996) *Darwin's Dangerous Idea: Evolution and the Meanings of Life*, Penguin Books, London. [A philosopher's view of natural selection.]

Richard Feynman (1990) *QED*, Penguin Books, London.

R. A. Fisher (1930, 1958) *The Genetical Theory of Natural Selection*, a variorum edition edited by H. Bennett (1999), Oxford University Press, Oxford. [This is primarily a book for specialists, and is not easy reading.]

Stephen W. Hawking (1988) *A Brief History of Time*, Bantam Press, London.

Fred Hoyle (1999) *Mathematics of Evolution*, Acorn Enterprises, Memphis, Tennessee. [This book is intended for specialists with some mathematical sophistication. A gentler account of Hoyle's views may be found in the next book in this list.]

Fred Hoyle and Chandra Wickramasinghe (1993) *Our Place in the Cosmos*, Phoenix, London.

Enrique Meléndez-Hevia (1993) *La Evolución del Metabolismo: hacia la Simplicidad*, Eudema, Madrid. [This book is written in Spanish, and is not available in English.]

T. H. Morgan (1932) *The Scientific Basis of Evolution*, W. W. Norton, New York.

Jacques Ninio (1982) *Molecular Approaches to Evolution*, Pitman Books, London.

Roger Penrose (1989) *The Emperor's New Mind*, Vintage, London.
Rober Rosen (1991) *Life Itself: a Comprehensive Inquiry into the Nature, Origin and Fabrication of Life*, Columbia University Press, New York.
Simon Singh (2000) *The Code Book*, Fourth Estate, London.
Adam Smith (1977) *An Inquiry into the Nature and Causes of the Wealth of Nations*, edited by Edwin Cannan, University of Chicago Press, Chicago. [A facsimile of the 1904 edition. Several other editions are available from other publishers.]
Lubert Stryer (1995) *Biochemistry*, 4th edn, Freeman, New York.
Steven Weinberg (1993) *Dreams of a Final Theory*, Vintage, London.

Index